ROBOT LEARNING

THE KLUWER INTERNATIONAL SERIES IN ENGINEERING AND COMPUTER SCIENCE

KNOWLEDGE REPRESENTATION, LEARNING AND EXPERT SYSTEMS
Consulting Editor
Tom Mitchell

Other books in the series:

ROBOT LEARNING

edited by

Jonalthan H. Connell
IBM, T.J. Watson Research Center

Sridhar Mahadevan
IBM, T.J. Watson Research Center

Kluwer Academic Publishers
Boston/Dordrecht/London

Distributors for North America:
Kluwer Academic Publishers
101 Philip Drive
Assinippi Park
Norwell, Massachusetts 02061 USA

Distributors for all other countries:
Kluwer Academic Publishers Group
Distribution Centre
Post Office Box 322
3300 AH Dordrecht, THE NETHERLANDS

Library of Congress Cataloging-in-Publication Data

A C.I.P. Catalogue record for this book is available from
the Library of Congress.

CONTENTS

CONTRIBUTORS

Christopher G. Atkeson
Artificial Intelligence Laboratory
Massachusetts Institute of Technology

Kenneth Basye
Department of Computer Science
Brown University

Rodney A. Brooks
Artificial Intelligence Laboratory
Massachusetts Institute of Technology

Jonathan H. Connell
T. J. Watson Research Center
International Business Machines

Thomas Dean
Department of Computer Science
Brown University

Richard Froom
Artificial Intelligence Laboratory
University of Texas at Austin

Leslie Kaelbling
Department of Computer Science
Brown University

Jonas Karlsson
Department of Computer Science
University of Rochester

Benjamin Kuipers
Artificial Intelligence Laboratory
University of Texas at Austin

Wan-Yik Lee
Artificial Intelligence Laboratory
University of Texas at Austin

Sridhar Mahadevan
T. J. Watson Research Center
International Business Machines

Maja J. Mataric
Artificial Intelligence Laboratory
Massachusetts Institute of Technology

Andrew W. Moore
Artificial Intelligence Laboratory
Massachusetts Institute of Technology

David Pierce
Artificial Intelligence Laboratory
University of Texas at Austin

Dean A. Pomerleau
School of Computer Science
Carnegie Mellon University

Josh Tenenberg
Department of Computer Science
University of Rochester

Steven Whitehead
Department of Computer Science
University of Rochester

PREFACE

Building a robot that learns to perform a task has been acknowledged as one of the major challenges facing artificial intelligence. Self-improving robots would relieve humans from much of the drudgery of programming and would potentially allow their operation in environments that were changeable or only partially known. Progress towards this goal would also make fundamental contributions to artificial intelligence by furthering our understanding of how to successfully integrate disparate abilities such as perception, planning, learning, and action.

Although its roots can be traced back to the late fifties, the area of robot learning has lately seen a resurgence of interest. The flurry of interest in robot learning has partly been fueled by exciting new work in the areas of reinforcement learning, behavior-based architectures, genetic algorithms, neural networks, and the study of artificial life. The renewed interest in robot learning has spawned many new research projects which have been discussed in an ever increasing number of workshop, conference, and journal papers. In the last two years, there were sessions entirely devoted to robot learning at the AAAI and NIPS conferences, and this year already shows promise of even more activity and interest in this area.

This book gives an overview of some of the current research projects in robot learning being carried out at leading universities and research laboratories across the country. The main research directions in robot learning covered in this book include: reinforcement learning, behavior-based architectures, neural networks, map learning, action models, navigation, and guided exploration. A bibliography of research papers and books is provided for further reading.

As one of the first books on robot learning, this book is intended to provide an easily accessible collection of papers on the subject. It may also be used as a supplementary textbook for a course in artificial intelligence, machine learning, or robotics. Given the interdiscplinary nature of the robot learning problem, the book may be of interest to a wide variety of readers, including computer scientists, roboticists, mechanical engineers, psychologists, ethologists, neurophysiologists, mathematicians, and philosophers.

The editors wish to thank all the chapter authors whose combined efforts made this book possible. In addition, Tom Mitchell provided valuable advice on preparing the book. We would also like to thank Carl Harris at Kluwer Academic Publishers for his encouragement in producing this book. Finally, we gratefully acknowledge the use of the computing facilities at IBM's T.J. Watson Research Center in the typesetting of this book.

ROBOT LEARNING

1

INTRODUCTION TO
ROBOT LEARNING

Jonathan H. Connell
Sridhar Mahadevan

*IBM T. J. Watson Research Center,
Box 704, Yorktown Heights, NY 10598*

ABSTRACT

In this chapter we provide an overview of the field of robot learning. We first discuss why robot learning is interesting and explain what is hard about it. We then characterize the robot learning problem and point out some major issues that need to be addressed. Next we survey some established techniques which are relevant to robot learning and then go on to review some of the recent research in robot learning. Finally, we briefly summarize the contents of the rest of the chapters in this book.

1 MOTIVATION

Building robots is hard and getting computers to learn is difficult. One might then ask, "Why do both when we currently don't understand either completely?" The answer is that we believe that some of the insights that result from studying robot learning may not be obtainable any other way. Robot learning forces us to deal with the issue of *integration* of multiple component technologies, such as sensing, planning, action, and learning. While one can study these components in isolation, studying them in the context of an actual robot can clarify issues arising in the interface between the various components. For example, a robot learning algorithm must be able to deal with the limitations of existing sensors.

Robot learning can be termed an *AI-complete* problem in the sense that it incorporates many hard issues in AI that may not be "solved" for some time to come. However, it is only by studying such problems that we can make incremental progress. Robot learning forces us to deal with the issue of *embedded systems* [Kaelbling 1990], where the learner is situated in an unknown dynamic environment. The issues that arise in robot learning are quite different from those that may arise in say a knowledge acquisition task for an expert system where, for example, the real time constraint may not be important.

If we view the ultimate goal of AI as bringing to reality systems such as the *R2D2* robot (from the movie *Star Wars*), then it is clear that we must study fundamental capabilities, such as learning, in the context of real robots (or sufficiently realistic simulated ones). Although we may not reach our ultimate destination for some time to come, there are many intermediate goals along the way which will offer ample reward in terms of useful practical systems and scientific insights.

Given that we are committed to making robots learn, the question now becomes, "Well, just what should they learn?". Generally speaking, we can distinguish three types of knowledge that it would be useful for a robot to automatically acquire.

- *Hard to Program Knowledge:* In any given task, we can usually distinguish between information that we can easily hardwire into the robot from that which would involve a lot of human effort. Consider the problem of getting a robot to operate a VCR designed for a human. It would be convenient to just show the robot how to operate the VCR by actually going through the motions ourselves. Clearly, this mode of teaching is much simpler and more effective than if we had to write a program for moving and coordinating the robot's joints, planning whole-arm trajectories, and specifying end-point compliances.

- *Unknown Information:* Sometimes the information necessary to program the robot is simply not readily available. For example, we might want to have a robot explore an unknown terrain, say on Mars or in the deep sea. Clearly, in such situations it is imperative that the robot be able to learn a map of the environment by exploring it. Or consider a generic household robot – the factory has programmed it to vacuum but it must learn for itself the layout of its buyer's home.

- *Changing Environments:* The world is a dynamic place. Objects move around from place to place, or appear and disappear. Even if we had a complete model of the environment to begin with, this knowledge could quickly become obsolete in a very dynamic environment. There are also slower changes, such as in the calibration of the robot's own sensors and effectors. Thus, it would be beneficial if a robot could constantly update its knowledge of both its internal and external environment.

However, because we are dealing with integrated, embodied systems there are a number of non-trivial real world issues that must be faced. Some of these are:

- *Sensor noise:* Most cheap-to-build robot sensors, such as sonar, are unreliable. Sonar transducers sometimes fail to see an object, or alternatively misjudge its distance. Thus, state descriptions computed from such sensors are bound to have inaccuracies in them, and some kind of probabilistic averaging is required.

- *Nondeterministic actions:* Since the robot has an incomplete model of its environment, actions will not always have similar effects. For example, if the robot picks up an object, the object may slip and fall sometimes, and other times the action may be successful. Planning becomes difficult because one has to allow for situations when a given action sequence fails to accomplish a goal.

- *Reactivity:* A robot must respond to unforeseen circumstances in real time. For example, a robot crossing a busy street intersection cannot afford the luxury of remaining motionless while it is computing the consequences of its current plan. In terms of learning, any reasonable algorithm must be tractable in that every step of the algorithm must terminate quickly.

- *Incrementality:* A robot has to collect the experience from which it is to learn the task. The data forming the experience is not available offline. The need for efficient exploration dictates that any reasonable learning algorithm must be incremental. Such algorithms should allow the robot to become better at deciding which part of the environment it needs to explore next.

■   *Limited training time:* Extended trials of multiple thousands of steps are very common in simulations, but are impractical for real robots. For a learning algorithm to be effective on a real robot, it must converge in a few thousand steps or less. Thus, the training time available on a real robot is very limited.

■   *Groundedness:* All the information that is available to a robot must come ultimately from its sensors (or be hardwired from the start). Since the state information is computed from sensors, a learning algorithm must be able to work with the limitations of perceptual devices. For example, a navigation robot may not be able to sense its exact coordinate location on a map. Simulation results that assume such exact location information are of limited value for real robots.

The chapters in this book address these constraints and illustrate how they influence the design of robot learning systems.

2   THE ROBOT LEARNING PROBLEM

Before examining specific techniques, we need to characterize the robot learning problem. We examine three aspects. First, we discuss where the training data comes from. Then we consider what types of feedback the robot might receive from the world to help it learn a task. Finally, we outline what sorts of knowledge need to be learned.

2.1   Training Data

Generally speaking, the robot learning problem is to infer a mapping from sensors to actions given a training sequence of sensory inputs, action outputs, and feedback values. An immediate question is where do these sequences come from? One possibility is that they are provided by a teacher. This situation corresponds to supervised learning. Here the robot is being passively guided through the task.

A more challenging and interesting situation arises when a robot attempts to learn a task in an unsupervised mode without the active guidance of a teacher. It is usually assumed here that the robot can recognize when it is performing the task properly (for example, reaching its goal). This learning situation is more challenging because the robot has to solve the task by executing trial-and-error actions thereby exploring the state space. It is more interesting than supervised learning because the robot has to collect its own training data. This flexibility opens up all kinds of interesting possibilities where the robot can actively choose to explore and experiment with certain parts of its environment. A robot can also bias the sample space of training data by selectively taking actions. Thus the robot can preferentially learn about the area around a known solution trajectory, or instead detect regions of state space in which it is ignorant and try to figure out how to obtain examples from there.

Supervised and unsupervised learning both play important roles in robot learning. Supervised training is useful because training time with real robots is rather limited. On the other hand, it is often difficult for a human to generate a rich enough sample space of training data. In such situations, trial-and-error unsupervised learning is more applicable. Before we can discuss these two types of learning more concretely, we need to understand the different types of feedback available to a robot. We turn to this topic next.

2.2 Feedback

For a robot to learn something, it must given some feedback concerning its performance. There are three primary types of feedback:

- *Scalar Feedback:* Here the robot is given a reward signal telling it how well it is currently doing at the task. In general, the sign of the reward indicates whether the robot's actions were good (positive reward) or bad (negative reward).

■ *Control Feedback:* The robot is directly told what action it is sup-
    posed to perform.

■ *Analytic Feedback:* In this arrangement the robot is given (or can
    itself generate) an explanation of why a certain action sequence will
    or will not achieve the goal.

Scalar feedback systems can be further categorized based on how often
they receive rewards. At one extreme, some robots receive no feedback
at all during most of their exploratory trial and error learning. For
example, the robot may be wandering from a start state to a goal state,
but only gets reinforcement when it actually reaches the goal [Sutton
1990]. Sometimes the only feedback occurs when the robot *fails* at its
task, such as in pole balancing [Barto *et al.* 1983]. A more informed
interaction occurs when the robot is given feedback prior to reaching
the goal. For example, the robot may be rewarded for reaching a critical
intermediate state in navigation tasks [Singh 1991]. Finally, we have
the situation where the robot receives scalar feedback at every step,
indicating the usefulness of the last action taken. One example is the
six-legged Genghis robot [Maes and Brooks 1990] which gets rewarded
by the amount it moved forward (measured by the number of turns of
a rear-mounted wheel). For navigation tasks, this would correspond to
giving a robot information regarding its current distance to the goal
[Pierce and Kuipers 1991].

Control feedback is essentially supervised learning. Typically the robot
is given a complete trajectory in the state space leading from an initial
state to the goal. For example, a human may guide a robot arm through
its motions [Asada and Yang 1989], or drive a robot vehicle through
an environment [Pomerleau 1989]. Supervised training demands more
effort from the teacher, but on the other hand potentially eliminates a
lot of unnecessary trial and error effort by the robot.

The easiest type of feedback to learn from is, of course, analytic feedback.
Typically the robot already possesses the knowledge necessary to solve
the task, but initially does so in a slow, lumbering fashion [Mitchell
1990]. To speed things up, the robot starts by generating an explanation
corresponding to a record of the problem-solving trace (say a search tree
of actions which includes a path to the goal). This explanation is then

cast in a form (say declarative) that makes it easy for the robot to determine its generality. The resulting macro-action helps accelerate the inference process of the robot controller in related situations.

It is important to note that our taxonomy of different types of feedback is somewhat oversimplified. In particular, there are intermediate situations between scalar and control feedback, or between control and analytic feedback. For example, the robot may be given gradient information to the goal [Pierce and Kuipers 1991], which is richer feedback than scalar, but is weaker than control feedback. An intermediate situation between control and analytic feedback is when the robot has a weak theory that can explain failures to achieve the goal, but not strong enough to plan to achieve the goal [Mitchell *et al.* 1989].

Of course, in a particular task a robot may also receive all three different types of feedback. It may initially operate completely in a trial and error mode, but get into a mode whereby it receives regular scalar rewards (say it is near a goal, and can sense how far away it is). Also, occasionally a teacher may be willing to tell the robot which action to perform [Clouse and Utgoff 1992]. Finally, the robot may be able to acquire a world model and use this model to speed up the learning of a task [Sutton 1990].

2.3   Credit Assignment

Given a particular mode of interaction with the environment, the next problem is how to learn from the sensor-action-feedback training sequences. This is the *credit assignment problem:* figure out what part of the robot's knowledge is defective, and how to fix it. Credit assignment is intimately related to generalization. There are so many possible combinations of sensory input, actions, and feedback values that it is unlikely that the robot will ever experience most of them. So how can it extrapolate what it has seen so far to similar situations which may arise in the future?

Figure 1 illustrates the nature of the credit assignment problem [Sutton 1984]. The dark curve depicts the time sequence of sensory states

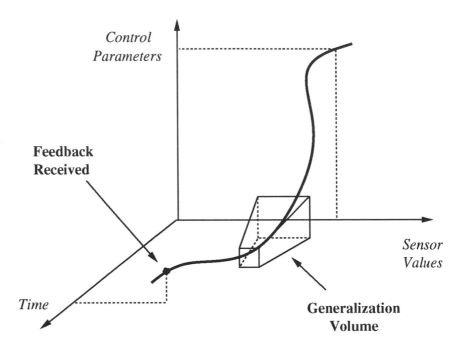

Figure 1.   The Credit Assignment Problem

perceived and the corresponding actions taken. At a certain point (the dark dot) the robot receives a reward. In this case the robot has to determine the extent of the generalization volume R such that it will still receive the same reward as long as the situation-action trajectory passes through R.

The *temporal* credit assignment problem refers to the task of figuring out which action or actions in the sequence was primarily responsible for the received reward (roughly, how long R is). The *structural* credit assignment problem, on the other hand, corresponds to figuring out the distribution of rewards across the state space (how wide R is). Given a different starting state, the robot needs to know the likelihood that the same action sequence will result in a similar outcome. Finally, there is the issue of control credit assignment (how high R is). Here the question is whether taking a slightly different action will still lead to the same reward.

Credit assignment can be viewed as the crux of the robot learning problem. There are many different techniques aimed at different parts of this problem. There are also a number of methods of incorporating bias into learners so that certain types of credit assignment become easier. The chapters in this book discuss a number of these points these in detail.

It turns out that temporal and structural credit assignment also provide a nice way to view the different types of environmental feedback. In the case of scalar feedback the robot has to solve both the structural and temporal credit assignment problems. Scalar feedback is so weak that there is no clear indication given as to what rewards the robot may receive in a state even slightly different from the one in the example sequence. Also, since rewards are often few and far between, the robot is faced with solving the full temporal credit assignment problem as well. Contrast this to the case of control feedback in which the robot is generally told precisely what to do at every step. Here the temporal credit assignment problem is largely solved by the teacher. Still, the robot must solve most of the structural credit assignment problem itself. If allowed to explore, it is likely to encounter novel states in which it has to determine the generality of the teacher's solution. Finally, in the case of analytic feedback, the robot is given considerable help on both the temporal and the structural credit assignment problems. A detailed explanation usually tells the robot not only what action to perform, but also helps the robot determine precisely the relevant range of situations.

3   BACKGROUND

In this section we survey some of the major methods for solving the structural and temporal credit assignment problem. The chapters in this book assume the reader already has some understanding of these methods. For those who are just entering the field this section provides capsule summaries of the relevant material.

3.1   Structural Credit Assignment

To a first approximation all the methods for solving this problem are essentially *function approximators* of one sort or another. They operate by imposing a similarity metric over the sensory input space and use this

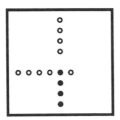  

Figure 2. The set of black and white dots (left) can be classified using a kd-tree (center) or a nearest-neighbor segmentation (right).

to interpolate between known control outputs. We can group the known methods into those that are *region-based* and those that are *point-based*.

In the region-based approach one divides up the input state space into a collection of non-overlapping areas. Each of these areas forms an equivalence class such that it is appropriate to take the same action for all states in the same area. Suppose, as shown in Figure 2, that there are a set of points (input instances) scattered on a two-dimensional plane (higher dimensions are possible) and that each point is labeled with the correct output response. Given a new input point the correct output (black or white in the figure) is just the value associated with whatever area that the point falls into.

One popular method for generating such areas is *kd-trees* [Friedman *et al.* 1977, Moore 1990]. The idea is to look at the statistical variance of the sample points along each of the input dimensions. We then choose the dimension with the largest variance and split the space with a line (or hyperplane for higher dimensions) at the median value along this axis. The original space has now been divided into two parts by an axis-parallel line. If all the points in a particular part have the same output value we stop here. Otherwise, for mixed values we again look at the variances of the remaining points along each of the dimensions and split each subregion independently. This continues until the system arrives at a number of rectangular regions each having only a single label.

Kd-trees generally ignore the labeling on the points. A set of techniques known as classification and regression trees (CART) overcome this limitation. These methods try to produce splits that maximize "purity" of the labels in each subregion. There are a number of ways to compute purity (see [Breiman *et al.* 1984]). A related technique from machine learning is ID3 [Quinlan 1983] which uses an information theoretic measure.

There are still other ways of dividing up the state space into regions. For instance, a linear classifier or perceptron [Minsky and Papert 1969] can be used to separate the space using arbitrary hyperplanes. Using conjunctions of such areas the space can be broken into a number of disjoint convex regions. Alternatively one could use a neural network to make boundaries with arbitrary shapes (see [Lippmann 1987] for a good review). A related technique is CMAC [Albus 1981] which uses a number of *overlapping* regions. This method computes the output value for some input as the average of the weights of all the regions containing the input.

A different approach to the function approximation problem is to use point-based instead of region-based classifiers. One popular technique is to generate an output response for a new point by interpolating the outputs of its *k nearest neighbors* (see [Moore 1990] for a survey).

Finally, instead of saving each data point, examples can often be summarized by forming *clusters* [Duda and Hart 1973]. These clusters are created by aggregating examples such that all the clusters have low variances and are reasonably separated from each other. The output for a new point is just the response associated with the nearest cluster center (or a value interpolated from several of them).

3.2 Temporal Credit Assignment

Most of the techniques for temporal credit assignment are variations on the *bucket brigade* method (cf. [Samuel 1959, Holland *et al.* 1986]). In these systems there is a sequence of "rule" firings which eventually leads to a reward. When this happens, the value of the last rule in the chain is increased due to the presumed imminence of a reward. The next time through the same sequence of rule firings, the penultimate rule will have its value increased because it immediately precedes a rule with a high value. Each time the system experiences this same sequence, the

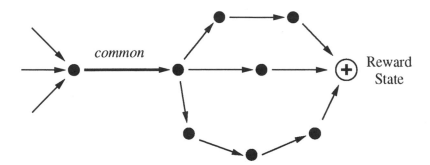

Figure 3. When multiple sequences of actions (arrows) lead to the same state it is the common action (dark) that is likely responsible for the reward.

value increases are gradually percolated further back through the chain of rules. If a number of the rules in a sequence do not directly influence the receipt of a reward then arbitrary substitutions can be made for them (see Figure 3). In this case only those rules which are at the confluence of multiple chains (i.e. that are common to all of them) end up receiving major increases in value. Q-learning [Watkins and Dayan 1992] and the method of Temporal Differences [Sutton 1988] formalize this idea.

The BOXES system [Mitchie and Chambers 1968, Barto 1987] is a commonly cited example of this method. Here the task was to continuously balance a free-swinging inverted pendulum (pole) affixed to a cart that could move to the left or the right along a track. The controller received a coarse estimate of the pole's angle and the cart's position along with the time derivatives of these quantities. From this information it had to decide in which direction (left or right) to apply a fixed force to the cart. The controller was punished when the pole fell over or the cart hit the end of its tracks. The problem is that the action taken right before the pole falls over is not the one truly responsible for the punishment received. It is really the action which first let the pole lean too far away from vertical that is the culprit.

A number of different techniques can be applied if the robot understands the context-dependent consequences of each of its actions. For instance, temporal credit assignment can be done directly using a dynamic programming search (see [Sedgewick 1983, Ross 1983]). Unfortunately, in most cases a suitable model is not available and, even when it is, the solution procedure is very time-consuming. Sutton's Dyna system [Sutton 1990] cleverly gets around these two problems by automatically building its own world model as it goes, and by using various approximations to full dynamic programming.

There are several other methods that can be used when one has a world model. If the model is symbolic then one use explanation based learning [Mitchell *et al.* 1986]. This solves not only the temporal but also the structural credit assignment problem. Suppose, instead, that both the world model (action-to-next-state-and-reward mapping) and the control policy (state-to-action mapping) are implemented as neural networks. These two networks can then be concatenated back-to-back to give a state-to-next-state mapping. Since this operation occurs over and over as time progresses, one can imagine an even larger network made out of sequential copies of this pair. Once this *unrolling in time* has been performed, a standard learning algorithm such as backpropagation can be used to simultaneously train both networks (cf. [Mitchell and Thrun 1992]).

4 DOMAINS

In terms of economic potential, the jobs most suited to robots are those that are *dirty, dull, or dangerous*. Robot learning is conceivably applicable in each of these areas. For instance, in a mining or toxic waste clean-up application the robot's joints might accumulate various sorts of sludge that could impede their motion. Having enough "compensatory plasticity" to adjust for this degradation would be desirable. As another example, imagine a system that learned how to assemble an item after watching a human do the task. Since robot hardware is more or less uniform, the program inferred could then be transferred to an army of assemblers. Finally, imagine building large structures in space, a very hazardous (and expensive) environment for humans to operate in. While teleoperation is sometimes practical, training an operator to handle all the details of such a robot can be time-consuming. In addition, the operation of such a complex mechanism quickly becomes fatiguing, especially when there are communication time lags that impair coordination. It would be much better if the robot could instead adapt to its user by, say, automatically inferring the next target site at the beginning of a motion. It might also learn to autonomously perform most of the dextrous manipulation required at the destination.

In terms of tasks, most of the chapters in this book are focussed on navigation in one form or another. However, there are several other tasks that are considered. These include walking, puck stacking, box pushing, and inferring sensory structure.

Many other robot learning tasks are not represented in this book. For instance, a lot of interesting work has been done with robot arms. Atkeson has developed a memory-based system which estimates the kinematic parameters of an arm from examples [Atkeson 1990b, Atkeson 1989]. After practicing a given motion for a while it is able to follow a specified trajectory much more closely. Miller's arm system learns to track objects using a wrist-mounted camera and a CMAC controller [Miller III 1986, Miller III et al. 1987]. The robot is initially given an approximate control law for moving the camera and then, through experience, builds a set of action-models that allows it to perform the task much more precisely and robustly. Another interesting robot learning problem involves hand-eye coordination. In Mel's MURPHY system the computer uses a neural network to determine the mapping between the joints of a real robot and the video image of these points [Mel 1990, Mel 1991]. With this learned mapping the robot can use mental imagery to automatically figure out how to reach for goals in its visual field despite intervening obstacles.

Another class of problems involves improving the performance of a robot at some task. Aboaf developed a robot which learns to throw a ball at a target [Aboaf 1988, Aboaf et al. 1988]. The computer refines the parameters of a supplied kinematic model based solely on whether the current throw was too high or too low. The system does not need to know the actual positions or velocities of the joints during the throws. This same technique also works for bouncing a ball on a paddle (sometimes called juggling). Another example is Asada's neural network-based robot which learns how to grind machine parts [Asada and Yang 1989, Asada 1990]. It passively observes an instrumented human perform the task and then generalizes this data to yield a suitable non-linear compliance function.

There has also been some work done on inferring symbolic procedures from sample execution traces. For instance, Dufay and Latombe built a system which planned part-mating motions and then replanned when unexpected events occurred [Dufay and Latombe 1984]. From these self-generated examples the system was able to construct a more robust generic plan. Andreae has a similar program in which a simulated robot learns to store cartons in neat aisles [Andreae 1985]. The system uses examples provided by a human, like where to put a box when the warehouse is empty and where to put the box when there are already four other boxes. From these it is able to generalize the parameters of certain motions, detect conditionals, and induce loops in the plan. Finally, Segre has a robotic system that uses a domain model in conjunction with explanation-based learning to perform assembly planning. Automatically generating an assembly plan involving a large number of separate motions, each of which can be parameterized in numerous ways, means searching a huge space. This system instead condenses a sequence of human-supplied motions into a effective new subassembly operator (to aid planning) and then further streamlines the resulting program based on causal dependencies [Segre 1988].

Some of the work in computer vision on object modeling and recognition is also relevant to robot learning. For instance, a robot which learns to put screwdrivers in one place and hammers in another must be able to distinguish between the two classes of objects. Some modeling systems start with gray scale images while others use depth maps obtained from stereo, shading, or a laser ranging (see [Ballard and Brown 1982] for examples). From this data the systems build up object models from primitives such as surface patches (e.g. [Vayda and Kak 1991]). The models generated range from semantic networks based on a generalized cylinder decomposition of objects [Connell and Brady 1987] to geometric hashing systems (e.g. [Bolle *et al.* 1989]).

Other systems use tactile feedback for object modeling. For instance, Stansfield built a system which used a vision system only to provide a coarse segmentation of an object [Stansfield 1988]. Using these cues a robotic touch sensor was directed to trace out the actual boundaries and investigate the surface properties of selected areas. The data from the two modalities was then combined to generate a rough 3-D description of the object. In another system a robot learns the shape of an object by repeatedly grasping it with a multi-fingered hand [Allen and Roberts 1989]. The contact points are then fit to a class of parameterized surface models called *super-quadrics*, a generalization of ellipsoids.

As can be seen, the potential applications of learning to robots are varied and diverse. A prerequisite to this endeavor, however, are testbed systems that can acquire potentially rich perceptual pictures of their world and interact smoothly with it. Robotic hardware and sensor technology is just getting the point where it is possible to build such devices and make them durable enough to run multiple experiments. Thus, we expect to see much more work in this field in the near future.

5   ROADMAP

We conclude this introduction to robot learning by giving the reader a brief roadmap to the rest of this book.

Chapter 2 by Pomerleau describes a neural network based system called ALVINN that learns to drive a real robot truck. Training data is provided by a human who simply drives the truck over the types of roads that it needs to be able to handle. One of the problems with using supervised learning for this task is that the robot seldom gets examples of what to do when it strays too far from the center of its lane. ALVINN solves this problem in an elegant manner. For every teacher provided example, ALVINN generates many more training instances by scaling and rotating the example sensory image.

Chapter 3 by Whitehead, Karlsson, and Tenenberg investigates the problem of scaling such reinforcement learning methods to tasks with multiple goals. The solution proposed here is to use a modular control architecture where each module learns to optimally solve a particular goal, and a central arbiter to select among modules. The authors also investigate combining a monolithic controller with a modular one to get a better tradeoff between optimality and accuracy for faster convergence

Chapter 4 by Moore and Atkeson introduces a new algorithm called *prioritized sweeping* for the prediction and control of stochastic systems.

This algorithm combines the real time nature of reinforcement learning and the efficient data usage of classical methods. The authors give several examples which demonstrate the superiority of this method in learning to optimally control Markovian decision systems.

Chapter 5 by Connell and Mahadevan addresses the limited training time available on a real robot and proposes four ways in which learning can be speeded up. These include breaking up a task into a set of subtasks, exploiting spatial locality, building action models to improve transfer of learning across tasks, and relying on a separate reactive controller to deal with minor changes in the environment. The authors show performance results using real robots on a box pushing task and a navigation task.

Chapter 6 by Kuipers, Froom, Lee, and Pierce develops a multi-level model of robot navigation learning. Their system determines the physical location of sensors and the primitive action types, extracts state variables and landmark locations, constructs topological maps, and then adds geometric information to flesh out the model. A number of results are shown based on a simulated robot and environment.

Chapter 7 by Dean, Bayse, and Kaelbling presents a theoretical analysis of the difficulty of learning a finite state model of a task environment (an idealization of navigation). Much of the previous work on this problem assumes perfect sensing whereas the authors consider not only imperfect sensors but also non-deterministic actions. Results are proved for special types of finite state environments and different degrees of stochasticity.

Finally, Chapter 8 by Brooks and Mataric discusses the general issue of scaling up learning to real robots. The authors propose four types of information that are useful for a robot to learn and survey some potential learning techniques, such as genetic algorithms. They also discuss the dangers of relying too much on simulated environments, and end by making a plea for more experimentation using real hardware.

2

KNOWLEDGE-BASED TRAINING OF ARTIFICIAL NEURAL NETWORKS FOR AUTONOMOUS ROBOT DRIVING

Dean A. Pomerleau

School of Computer Science,
Carnegie Mellon University, Pittsburgh, PA 15213

ABSTRACT

Many real world problems require a degree of flexibility that is difficult to achieve using hand programmed algorithms. One such domain is vision-based autonomous driving. In this task, the dual challenges of a constantly changing environment coupled with a real time processing constrain make the flexibility and efficiency of a machine learning system essential. This chapter describes just such a learning system, called ALVINN (Autonomous Land Vehicle In a Neural Network). It presents the neural network architecture and training techniques that allow ALVINN to drive in a variety of circumstances including single-lane paved and unpaved roads, multilane lined and unlined roads, and obstacle-ridden on- and off-road environments, at speeds of up to 55 miles per hour.

1 INTRODUCTION

Autonomous navigation is a difficult problem for traditional vision and robotic techniques, primarily because of the noise and variability associated with real world scenes. Autonomous navigation systems based on traditional image processing and pattern recognition techniques often perform well under certain conditions but have problems with others. Part of the difficulty stems from the fact that the processing performed by these systems remains fixed across various environments.

Figure 1. The CMU Navlab Autonomous Navigation Testbed

Artificial neural networks have displayed promising performance and flexibility in other domains characterized by high degrees of noise and variability, such as handwritten character recognition [LeCun *et al.* 1989] and speech recognition [Waibel *et al.* 1988] and face recognition [Cottrell 1990]. ALVINN (Autonomous Land Vehicle In a Neural Network) is a system that brings the flexibility of connectionist learning techniques to the task of autonomous robot navigation. Specifically, ALVINN is an artificial neural network designed to control the Navlab II, Carnegie Mellon's autonomous driving test vehicle (See Figure 1).

This chapter describes the architecture, training and performance of the ALVINN system. It demonstrates how simple connectionist networks can learn to precisely guide a mobile robot in a wide variety of situations when trained appropriately. In particular, this chapter presents training techniques that allow ALVINN to learn in under 5 minutes to autonomously control the Navlab by watching a human driver's response to new situations. Using these techniques, ALVINN has been trained to drive in a variety of circumstances including single-lane paved and unpaved roads, multilane lined and unlined roads, and obstacle-ridden on- and off-road environments, at speeds of up to 55 miles per hour.

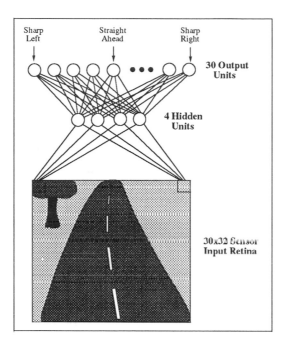

Figure 2. Neural network architecture for autonomous driving.

2 NETWORK ARCHITECTURE

The basic network architecture employed in the ALVINN system is a single hidden layer feedforward neural network (See Figure 2). The input layer now consists of a single 30x32 unit "retina" onto which a sensor image from either a video camera or a scanning laser rangefinder is projected. Each of the 960 input units is fully connected to the hidden layer of 4 units, which is in turn fully connected to the output layer. The 30 unit output layer is a linear representation of the currently appropriate steering direction which may serve to keep the vehicle on the road or to prevent it from colliding with nearby obstacles[1]. The centermost output unit represents the "travel straight ahead" condition, while units to the left and right of center represent successively sharper left and right turns. The units on the extreme left and right of the output vector represent turns with a 20m radius to the left and right respectively, and the units

---

[1]The task a particular driving network performs depends on the type of input sensor image and the driving situation it has been trained to handle.

in between represent turns which decrease linearly in their curvature down to the "straight ahead" middle unit in the output vector.

To drive the Navlab, an image from the appropriate sensor is reduced to 30x32 pixels and projected onto the input layer. After propagating activation through the network, the output layer's activation profile is translated into a vehicle steering command. The steering direction dictated by the network is taken to be the center of mass of the "hill" of activation surrounding the output unit with the highest activation level. Using the center of mass of activation instead of the most active output unit when determining the direction to steer permits finer steering corrections, thus improving ALVINN's driving accuracy.

3   NETWORK TRAINING

The network is trained to produce the correct steering direction using the backpropagation learning algorithm [Rumelhart *et al.* 1986]. In backpropagation, the network is first presented with an input and activation is propagated forward through the network to determine the network's response. The network's response is then compared with the known correct response. If the network's actual response does not match the correct response, the weights between connections in the network are modified slightly to produce a response more closely matching the correct response.

Autonomous driving has the potential to be an ideal domain for a supervised learning algorithm like backpropagation since there is a readily available teaching signal or "correct response" in the form of the human driver's current steering direction. In theory it should be possible to teach a network to imitate a person as they drive using the current sensor image as input and the person's current steering direction as the desired output. This idea of training "on-the-fly" is depicted in Figure 3.

Training on real images would dramatically reduce the human effort required to develop networks for new situations, by eliminating the need for a hand-programmed training example generator. On-the-fly training should also allow the system to adapt quickly to new situations.

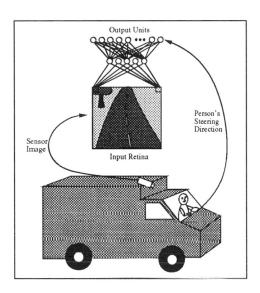

Figure 3. Schematic representation of training "on the fly". The network is shown images from the onboard sensor and trained to steer in the same direction as the human driver.

3.1   Potential Problems

There are two potential problems associated with training a network using live sensor images as a person drives. First, since the person steers the vehicle down the center of the road during training, the network will never be presented with situations where it must recover from misalignment errors. When driving for itself, the network may occasionally stray from the road center, so it must be prepared to recover by steering the vehicle back to the middle of the road. The second problem is that naively training the network with only the current video image and steering direction may cause it to overlearn recent inputs. If the person drives the Navlab down a stretch of straight road at the end of training, the network will be presented with a long sequence of similar images. This sustained lack of diversity in the training set will cause the network to "forget" what it had learned about driving on curved roads and instead learn to always steer straight ahead.

Both problems associated with training on-the-fly stem from the fact that back-propagation requires training data which is representative of

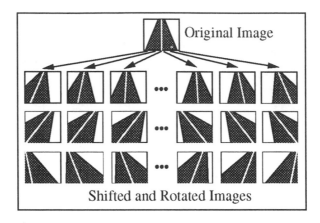

Figure 4. The single original video image is shifted and rotated to create multiple training exemplars in which the vehicle appears to be at different locations relative to the road.

the full task to be learned. The first approach we considered for increasing the training set diversity was to have the driver swerve the vehicle during training. The idea was to teach the network how to recover from mistakes by showing it examples of the person steering the vehicle back to the road center. However this approach was deemed impractical for two reasons. First, training while the driver swerves would require turning learning off while the driver steers the vehicle off the road, and then back on when he swerves back to the road center. Without this ability to toggle the state of learning, the network would incorrectly learn to imitate the person swerving off the road as well as back on. While possible, turning learning on and off would require substantial manual input during the training process, which we wanted to avoid. The second problem with training by swerving is that it would require swerving in many circumstances to enable the network to learn a general representation. This would be time consuming, and also dangerous when training in traffic.

3.2   Solution - Transform the Sensor Image

To achieve sufficient diversity of real sensor images in the training set, without the problems associated with training by swerving, we have developed a technique for transforming sensor images to create additional training exemplars. Instead of presenting the network with only the current sensor image and steering direction, each sensor image is shifted and rotated in software to create additional images in which the vehicle appears to be situated differently relative to the environment (See Figure 4). The sensor's position and orientation relative to the ground plane are known, so precise transformations can be achieved using perspective geometry.

The image transformation is performed by first determining the area of the ground plane which is visible in the original image, and the area that should be visible in the transformed image. These areas form two overlapping trapezoids as illustrated by the aerial view in Figure 5. To determine the appropriate value for a pixel in the transformed image, that pixel is projected onto the ground plane, and then back-projected into the original image. The value of the corresponding pixel in the original image is used as the value for the pixel in the transformed image. One important thing to realize is that the pixel-to-pixel mapping which implements a particular transformation is constant. In other words, assuming a planar world, the pixels which need to be sampled in the original image in order to achieve a specific shift and translation in the transformed image always remain the same. In the actual implementation of the image transformation technique, ALVINN takes advantage of this fact by precomputing the pixels that need to be sampled in order to perform the desired shifts and translations. As a result, transforming the original image to change the apparent position of the vehicle simply involves changing the pixel sampling pattern during the image reduction phase of preprocessing. Therefore, creating a transformed low resolution image takes no more time than is required to reduce the image resolution to that required by the ALVINN network. Obviously the environment is not always flat. But the elevation changes due to hills or dips in the road are small enough so as not to significantly violate the planar world assumption.

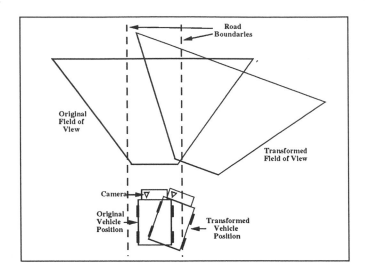

Figure 5. An aerial view of the vehicle at two different positions, with the corresponding sensor fields of view. To simulate the image transformation that would result from such a change in position and orientation of the vehicle, the overlap between the two field of view trapezoids is computed and used to direct resampling of the original image.

Extrapolating Missing Pixels

The less than complete overlap between the trapezoids of Figure 5 illustrates the need for one additional step in the image transformation scheme. The extra step involves determining values for pixels which have no corresponding pixel in the original image. Consider the transformation illustrated in Figure 6. To make it appear that the vehicle is situated one meter to the right of its position in the original image requires not only shifting pixels in the original image to the left, but also filling in the unknown pixels along the right edge. Notice the number of pixels per row whose value needs to be extrapolated is greater near the bottom of the image than at the top. This is because the one meter of unknown ground plane to the right of the visible boundary in the original image covers more pixels at the bottom than at the top. We have experimented with two techniques for extrapolating values for these unknown pixels (See Figure 7).

In the first technique, to determine the value for a pixel that projects to the ground plane at point A in the transformed image, the closest ground plane point in the original viewing trapezoid (point B) is found. This point is then back-projected into the original image to find the appropriate pixel to sample. The image in the top right shows the sampling performed to fill in the missing pixel using this extrapolation scheme. The problem with this technique is that it results in the "smearing" of the image approximately along rows of the image, as illustrated in the middle image of Figure 8. In this figure, the leftmost image represents an actual reduced resolution image of a two-lane road coming from the camera. Notice the painted lines delineating the center and right boundaries of the lane. The middle image shows the original image transformed to make it appear that the vehicle is one meter to the right of its original position using the extrapolation technique described above. The line down the right side of the road can be seen smearing to the right where it intersects the border of the original image. Because the length of this smear is highly correlated with the correct steering direction, the network learns to depend on the size of this smear to predict the correct steering direction. When driving on its own however, this lateral smearing of features is not present, so the network performs poorly.

To eliminate this artifact of the transformation process, we implemented a more realistic extrapolation technique which relies on the fact that interesting features (like road edges and painted lane markers) normally run parallel to the road, and hence parallel to the vehicle's current direction. With this assumption, to extrapolate a value for the unknown pixel A in Figure 7, the appropriate ground plane point to sample from the original image's viewing trapezoid is not the closest point (point B), but the nearest point in the original image's viewing trapezoid along the line that runs through point A and is parallel to the vehicle's original heading (point C).

The effect this improved extrapolation technique has on the transformed image can be seen schematically in the bottom image on the right of Figure 7. This technique results in extrapolation along the line connecting a missing pixel to the vanishing point, as illustrated in the lower right image. The realism advantage this extrapolation technique has over the previous scheme can be seen by comparing the image on the right of

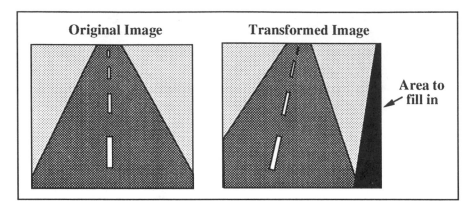

Figure 6. A schematic example of an original image, and a transformed image in which the vehicle appears one meter to the right of its initial position. The black region on the right of the transformed image corresponds to an unseen area in the original image. These pixels must be extrapolated from the information in the original image.

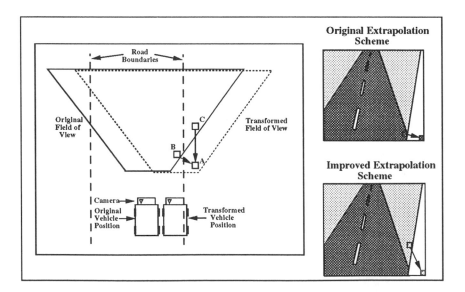

Figure 7. An aerial view (left) and image based view (right) of the two techniques used to extrapolate the values for unknown pixels. See text for explanation.

Figure 8. Three reduced resolution images of a two-lane road with lines painted down the middle and right side. The left image is the original coming directly from the camera. The middle image was created by shifting the original image to make it appear the vehicle was situated one meter to the right of its original position using the first extrapolation technique described in the text. The right image shows the same shift of the original image, but using the more realistic extrapolation technique.

Figure 8 with the middle image. The line delineating the right side of the lane, which was unrealistically smeared using the previous method, is smoothly extended in the image on the right, which was created by shifting the original image by the same amount as in the middle image, but using the improved extrapolation method.

The improved transformation scheme certainly makes the transformed images look more realistic, but to test whether it improves the network's driving performance, we did the following experiment. We first collected actual two-lane road images like the one shown on the left side of Figure 8 along with the direction the driver was steering when the images were taken. We then trained two networks on this set of images. The first network was trained using the naive transformation scheme and the second using the improved transformation scheme. The magnitude of the shifts and rotations, along with the buffering scheme used in the training process are described in detail below. The networks were then tested on a disjoint set of real two-lane road images, and the steering direction dictated by the networks was compared with the person's steering direction on those images. The network trained using the more realistic transformation scheme exhibited 37% less steering error on the 100 test images than the network trained using the naive transformation scheme. In more detail, the amount of steering error a network produces is measured as the distance, in number of units (i.e. neurons), between

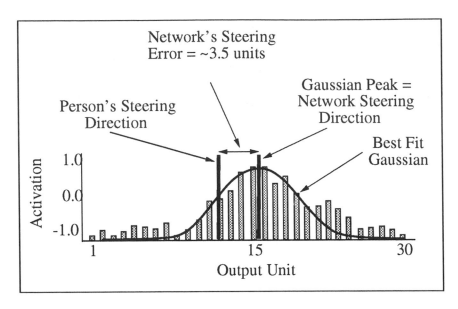

Figure 9. To calculate a network's steering error the best fit gaussian is found to the network's output activation profile. The distance between the peak of the best fit gaussian and the position in the output vector representing the reference steering direction (in this case the person's steering direction) is calculated. This distance, measured in units or neurons between the two positions, is defined to be the network's steering error.

the peak of the network's "hill" of activation in the output vector and the "correct" position, in this case the direction the person was actually steering in. This steering error measurement is illustrated in Figure 9. In this case, the network trained with the naive transformation technique had an average steering error across the 100 test images of 3.5 units, while the network trained with the realistic transformations technique had an average steering error of only 2.2 units.

3.3   Transforming the Steering Direction

As important as the technique for transforming the input images is the method used to determine the correct steering direction for each of the transformed images. The correct steering direction as dictated by the driver for the original image must be altered for each of the transformed images to account for the altered vehicle placement. This is done using a simple model called pure pursuit steering [Wallace *et al.* 1985]. In the pure pursuit model, the "correct" steering direction is the one that will bring the vehicle to a desired location (usually the center of the road) a fixed distance ahead. The idea underlying pure pursuit steering is illustrated in Figure 10. With the vehicle at position A, driving for a predetermined distance along the person's current steering arc would bring the vehicle to a "target" point T, which is assumed to be in the center of the road.

After transforming the image with a horizontal shift s and rotation θ to make it appear that the vehicle is at point B, the appropriate steering direction according to the pure pursuit model would also bring the vehicle to the target point T. Mathematically, the formula to compute the radius of the steering arc that will take the vehicle from point B to point T is

$$r = \frac{l^2 + d^2}{2d}$$

where r is the steering radius l is the lookahead distance and d is the distance from point T the vehicle would end up at if driven straight ahead from point B for distance l. The displacement d can be determined using the following formula:

$$d = \cos\theta \cdot (d_p + s + l\tan\theta)$$

where d_p is the distance from point T the vehicle would end up if it drove straight ahead from point A for the lookahead distance l, s is the horizontal distance from point A to B, and θ is the vehicle rotation from point A to B. The quantity d_p can be calculated using the following equation:

$$d_p = r_p - \sqrt{r_p^2 - l^2}$$

where r_p is the radius of the arc the person was steering along when the image was taken.

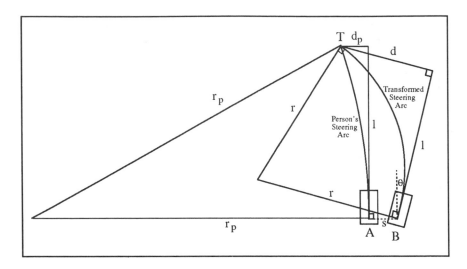

Figure 10.   Illustration of the "pure pursuit" model of steering. See text for explanation.

The only remaining unspecified parameter in the pure pursuit model is l, the distance ahead of the vehicle to select a point to steer towards. Empirically, I have found that over the speed range of 5 to 55 mph, accurate and stable vehicle control can be achieved using the following rule: look ahead the distance the vehicle will travel in 2-3 seconds.

Interestingly, with this empirically determined rule for choosing the lookahead distance, the pure pursuit model of steering is a fairly good approximation to how people actually steer. Reid, Solowka and Billing [Reid *et al.* 1981] found that at 50km/h, human subjects responded to a 1m lateral vehicle displacement with a steering radius ranging from 511m to 1194m. With a lookahead equal to the distance the vehicle will travel in 2.3 seconds, the pure pursuit model dictates a steering radius of 594m, within the range of human responses. Similarly, human subjects reacted to a 1 degree heading error relative to the current road direction with a steering radius ranging from 719m to 970m. Again using the 2.3 second travel distance for lookahead, the pure pursuit steering model's dictated radius of 945m falls within the range of human responses.

Like the image transformation scheme, the steering direction transformation technique uses a simple model to determine how a change in the vehicle's position and/or orientation would affect the situation. In the image transformation scheme, a planar world hypothesis and rules of perspective projection are used to determine how changing the vehicle's position and/or orientation would affect the sensor image of the scene ahead of the vehicle. In the steering direction transformation technique, a model of how people drive is used to determine how a particular vehicle transformation should alter the correct steering direction. In both cases, the transformation techniques are independent of the driving situation. The person could be driving on a single lane dirt road or a multi lane highway: the transformation techniques would be the same.

Anthropomorphically speaking, transforming the sensor image to create more training images is equivalent to telling the network "I don't know what features in the image are important for determining the correct direction to steer, but whatever they are, here are some other positions and orientations you may see them in". Similarly, the technique for transforming the steering direction for each of these new training images is equivalent to telling the network "whatever the important features are, if you see them in this new position and orientation, here is how your response should change". Because it does not rely on a strong model of what important image features look like, but instead acquires this knowledge through training, the system is able to drive in a wide variety of circumstances, as will be seen later in the chapter.

These weak models are enough to solve the two problems associated with training in real time on sensor data. Specifically, using transformed training patterns allows the network to learn how to recover from driving mistakes that it would not otherwise encounter as the person drives. Also, overtraining on repetitive images is less of a problem, since the transformed training exemplars maintain variety in the training set.

3.4 Adding Diversity Through Buffering

As additional insurance against the effects of repetitive exemplars, the training set diversity is further increased by maintaining a buffer of previously encountered training patterns. When new training patterns are acquired through digitizing and transforming the current sensor image, they are added to the buffer, while older patterns are removed. We have experimented with four techniques for determining which patterns to replace. The first is to replace oldest patterns first. Using this scheme, the training pattern buffer represents a history of the driving situations encountered recently. But if the driving situation remains unchanged for a period of time, such as during an extended right turn, the buffer will loose its diversity and become filled with right turn patterns. The second technique is to randomly choose old patterns to be replaced by new ones. Using this technique, the laws of probability help ensure somewhat more diversity than the oldest pattern replacement scheme, but the buffer will still become biased during monotonous stretches.

The next solution we developed to encourage diversity in the training was to replace those patterns on which the network was making the lowest error, as measured by the sum squared difference between the network's output and the desired output. The idea was to eliminate the patterns the network was performing best on, and leave in the training set those images the network was still having trouble with. The problem with this technique results from the fact that the human driver doesn't *always* steer in the correct direction. Occasionally he may have a lapse of attention for a moment and steer in an incorrect direction for the current situation. If a training exemplar was collected during this momentary lapse, under this replacement scheme it will remain there in the training buffer for a long time, since the network will have trouble outputting a steering response to match the person's incorrect steering command. In fact, using this replacement technique, the only way the pattern would be removed from the training set would be if the network learned to duplicate the incorrect steering response, obviously not a desired outcome. I considered replacing both the patterns with the lowest error *and* the patterns with the highest error, but decided against it since high network error on a pattern might also result on novel input image with a correct response associated with it. A better method to eliminate this problem is to add a random replacement probability to all patterns in the training buffer. This ensured that even if the network never learns to produce the same steering response as the person on an image, that image will eventually be eliminated from the training set.

While this augmented lowest-error-replacement technique did a reasonable job of maintaining diversity in the training set, we found a more straightforward way of accomplishing the same result. To make sure the buffer of training patterns does not become biased towards one steering direction, we add a constraint to ensure that the mean steering direction of all the patterns in the buffer is as close to straight ahead as possible. When choosing the pattern to replace, I select the pattern whose replacement will bring the average steering direction closest to straight. For instance, if the training pattern buffer had more right turns than left, and a left turn image was just collected, one of the right turn images in the buffer would be chosen for replacement to move the average steering direction towards straight ahead. If the buffer already had a straight ahead average steering direction, then an old pattern requiring a similar steering direction the new one would be replaced in order to maintain the buffer's unbiased nature. By actively compensating for steering bias in the training buffer, the network never learns to consistently favor one steering direction over another. This active bias compensation is a way to build into the network a known constraint about steering: in the long run right and left turns occur with equal frequency.

3.5   Training Details

The final details required to specify the training on-the-fly process are the number and magnitude of transformations to use for training the network. The following quantities have been determined empirically to provide sufficient diversity to allow networks to learn to drive in a wide variety of situations. The original sensor image is shifted and rotated 14 times using the technique describe above to create 14 training exemplars. The size of the shift for each of the transformed exemplars is chosen randomly from the range -0.6 to +0.6 meters, and the amount of rotation is chosen from the range -6.0 to +6.0 degrees. In the image formed by the camera on the Navlab, which has a 42 degree horizontal field of view, an image with a maximum shift of 0.6m results in the road shifting approximately 1/3 of the way across the input image at the bottom.

Before the randomly selected shift and rotation is performed on the original image, the steering direction that would be appropriate for the resulting transformed image is computed using the formulas given above. If the resulting steering direction is sharper than the sharpest turn representable by the network's output (usually a turn with a 20m radius), then the transformation is disallowed and a new shift distance and rotation magnitude are randomly chosen. By eliminating extreme and unlikely conditions from the training set, such as when the road is shifted far to the right and vehicle is heading sharply to the left, the network is able to devote more of its representation capability to handling plausible scenarios.

The 14 transformed training patterns, along with the single pattern created by pairing the current sensor image with the current steering direction, are inserted into the buffer of 200 patterns using the replacement strategy described above. After this replacement process, one forward and one backward pass of the backpropagation algorithm is performed on the 200 exemplars to update the network's weights, using a learning rate of 0.01 and a momentum of 0.8. The entire process is then repeated. Each cycle requires approximately 2.5 seconds on the three Sun Sparcstations onboard the vehicle. One of the Sparcstation performs the sensor image acquisition and preprocessing, the second implements the neural network simulation, and the third takes care of communicating with the vehicle controller and displaying system parameters for the human observer. The network requires approximately 100 iterations through this digitize-replace-train cycle to learn to drive in the domains that have been tested. At 2.5 seconds per cycle, training takes approximately four minutes of human driving over a sample stretch of road. During the training phase, the person drives at approximately the speed at which the network will be tested, which ranges from 5 to 55 miles per hour.

4   PERFORMANCE IMPROVEMENT USING TRANSFORMATIONS

The performance advantage this technique of transforming and buffering training patterns offers over the more naive methods of training on real sensor data is illustrated in Figure 11. This graph shows the vehicle's displacement from the road center measured as three different networks drove at 4 mph over a 100 meter section of a single lane paved bike path which included a straight stretch and turns to the left and right. The three networks were trained over a 150 meter stretch of the path which

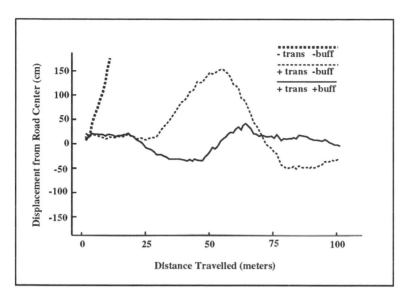

Figure 11. Vehicle displacement from the road center as the Navlab was driven by networks trained using three different techniques.

was disjoint from the test section and which ended in an extended right turn.

The first network, labeled "-trans -buff", was trained using just the images coming from the video camera. That is, during the training phase, an image was digitized from the camera and fed into the network. One forward and backward pass of back-propagation was performed on that training exemplar, and then the process was repeated. The second network, labeled "+trans -buff", was trained using the following technique. An image was digitized from the camera and then transformed 14 times to create 15 new training patterns as described above. A forward and backwards pass of back-propagation was then performed on each of these 15 training patterns and then the process was repeated. The third network, labeled "+trans +buff" was trained using the same transformation scheme as the second network, but with the addition of the image buffering technique described above to prevent overtraining on recent images.

Note that all three networks were presented with the same number of images. The transformation and buffering schemes did not influence the quantity of data the networks were trained on, only its distribution.

The "-trans -buff" network was trained on closely spaced actual video images. The "+trans -buff" network was presented with 15 times fewer actual images, but its training set also contained 14 transformed images for every "real" one. The "+trans +buff" network collected even fewer live images, since it performed a forward and backward pass through its buffer of 200 patterns before digitizing a new one.

The accuracy of each of the three networks was determined by manually measuring the vehicle's lateral displacement relative to the road center as each network drove. The network trained on only the current video image quickly drove off the right side of the road, as indicated by its rapidly increasing displacement from the road center. The problem was that the network overlearned the right turn at the end of training and became biased towards turning right. Because of the increased diversity provided by the image transformation scheme, the second network performed much better than the first. It was able to follow the entire test stretch of road. However it still had a tendency to steer too much to the right, as illustrated in the graph by the vehicle's positive displacement over most of the test run. In fact, the mean position of the vehicle was 28.9cm right of the road center during the test. The variability of the errors made by this network was also quite large, as illustrated by the wide range of vehicle displacement in the "+trans -buff" graph. Quantitatively, the standard deviation of this network's displacement was 62.7cm.

The addition of buffering previously encountered training patterns eliminated the right bias in the third network, and also greatly reduced the magnitude of the vehicle's displacement from the road center, as evidenced by the "+trans +buff" graph. While the third network drove, the average position of the vehicle was 2.7cm right of center, with a standard deviation of only 14.8cm. This represents a 423% improvement in driving accuracy.

A separate test was performed to compare the steering accuracy of the network trained using both transformations and buffering with the steering accuracy of a human driver. This test was performed over the same stretch of road as the previous one, however the road was less obscured by fallen leaves in this test, resulting in better network performance. Over three runs, with the network driving at 5 miles per hour along the

100 meter test section of road, the average position of the vehicle was 1.6cm right of center, with a standard deviation of 7.2cm. Under human control, the average position of the vehicle was 4.0cm right of center, with a standard deviation of 5.47cm. The average distance the vehicle was from the road center while the person drove was 5.70cm. It appears that the human driver, while more consistent than the network, had an inaccurate estimate of the vehicle's centerline, and therefore drove slightly right of the road center. Studies of human driving performance have found similar steady state errors and variances in vehicle lateral position. Blaauw [Blaauw 1982] found consistent displacements of up to 7cm were not uncommon when people drove on highways. Also for highway driving, Blaauw reports standard deviations in lateral error up to 16.6cm.

5   RESULTS AND COMPARISON

The competence of the ALVINN system is also demonstrate by the range of situations in which it has successfully driven.

The training on-the-fly scheme gives ALVINN a flexibility which is novel among autonomous navigation systems. It has allowed me to successfully train individual networks to drive in a variety of situations, including a single-lane dirt access road, a single-lane paved bicycle path, a two-lane suburban neighborhood street, and a lined two-lane highway (See Figure 12). Using other sensor modalities as input, including laser range images and laser reflectance images, individual ALVINN networks have been trained to follow roads in total darkness, to avoid collisions in obstacle rich environments, and to follow alongside railroad tracks. ALVINN networks have driven without intervention for distances of up to 22 miles. In addition, since determining the steering direction from the input image merely involves a forward sweep through the network, the system is able to process 15 images per second, allowing it to drive at up to 55 miles per hour. This is over four times faster than any other sensor-based autonomous system using the same processing hardware, has driven the Navlab [Crisman and Thorpe 1990, Kluge and Thorpe 1990].

Figure 12. Video images taken on three of the road types ALVINN modules have been trained to handle. They are, from left to right, a single-lane dirt access road, a single-lane paved bicycle path, and a lined two-lane highway.

The level of flexibility across driving situations exhibited by ALVINN would be difficult to achieve without learning. It would require the programmer to 1) determine what features are important for the particular driving domain, 2) program detectors (using statistical or symbolic techniques) for finding these important features and 3) develop an algorithm for determining which direction to steer from the location of the detected features. As a result, while hand programmed systems have been developed to drive in some of the individual domains ALVINN can handle [Crisman and Thorpe 1990, Kluge and Thorpe 1990, Turk *et al.* 1988, Dickmanns and Zapp 1987], none have duplicated ALVINN's flexibility.

ALVINN is able to *learn* for each new domain what image features are important, how to detect them and how to use their position to steer the vehicle. Analysis of the hidden unit representations developed in different driving situations shows that the network forms detectors for the image features which correlate with the correct steering direction. When trained on multi-lane roads, the network develops hidden unit feature detectors for the lines painted on the road, while in single-lane driving situations, the detectors developed are sensitive to road edges and road-shaped regions of similar intensity in the image. For a more detailed analysis of ALVINN's internal representations see [Pomerleau 1989, Pomerleau 1990].

This ability to utilize arbitrary image features can be problematic. This was the case when ALVINN was trained to drive on a poorly defined dirt road with a distinct ditch on its right side. The network had no problem learning and then driving autonomously in one direction, but when driving the other way, the network was erratic, swerving from one side of the road to the other. After analyzing the network's hidden representation, the reason for its difficulty became clear. Because of the poor distinction between the road and the non-road, the network had developed only weak detectors for the road itself and instead relied heavily on the position of the ditch to determine the direction to steer. When tested in the opposite direction, the network was able to keep the vehicle on the road using its weak road detectors but was unstable because the ditch it had learned to look for on the right side was now on the left. Individual ALVINN networks have a tendency to rely on *any* image feature consistently correlated with the correct steering direction. Therefore, it is important to expose them to a wide enough variety of situations during training so as to minimize the effects of transient image features.

On the other hand, experience has shown that it is more efficient to train several domain specific networks for circumstances like one-lane vs. two-lane driving, instead training a single network for all situations. To prevent this network specificity from reducing ALVINN's generality, we are currently implementing connectionist and non-connectionist techniques for combining networks trained for different driving situations. Using a simple rule-based priority system similar to the subsumption architecture [Brooks 1986], we have combined a road following network and an obstacle avoidance network. The road following network uses video camera input to follow a single-lane road. The obstacle avoidance network uses laser rangefinder images as input. It is trained to swerve appropriately to prevent a collision when confronted with obstacles and to drive straight when the terrain ahead is free of obstructions. The arbitration rule gives priority to the road following network when determining the steering direction, except when the obstacle avoidance network outputs a sharp steering command. In this case, the urgency of avoiding an imminent collision takes precedence over road following and the steering direction is determined by the obstacle avoidance network. Together, the two networks and the arbitration rule comprise a system capable of staying on the road and swerving to prevent collisions.

To facilitate other rule-based arbitration techniques, we have adding to ALVINN a non-connectionist module which maintains the vehicle's position on a map [Pomerleau *et al.* 1991]. Knowing its map position allows ALVINN to use arbitration rules such as "when on a stretch of two lane highway, rely primarily on the two lane highway network". This symbolic mapping module also allows ALVINN to make high level, goal-oriented decisions such as which way to turn at intersections and when to stop at a predetermined destination.

Finally, we are experimenting with connectionist techniques, such as the task decomposition architecture [Jacobs *et al.* 1990] and the meta-pi architecture [Hampshire and Waibel 1989], for combining networks more seamlessly than is possible with symbolic rules. These connectionist arbitration techniques will enable ALVINN to combine outputs from networks trained to perform the same task using different sensor modalities and to decide when a new expert must be trained to handle the current situation.

6   DISCUSSION

A truly autonomous mobile vehicle must cope with a wide variety of driving situations and environmental conditions. As a result, it is crucial that an autonomous navigation system possess the ability to adapt to novel domains. Supervised training of a connectionist network is one means of achieving this adaptability. But teaching an artificial neural network to drive based on a person's driving behavior presents a number of challenges. Prominent among these is the need to maintain sufficient variety in the training set to ensure that the network develops a sufficiently general representation of the task. Two characteristics of real sensor data collected as a person drives which make training set variety difficult to maintain are temporal correlations and the limited range of situations encountered. Extended intervals of nearly identical sensor input can bias a network's internal representation and reduce later driving accuracy. The human trainer's high degree of driving accuracy severely restricts the variety of situations covered by the raw sensor data.

The techniques for training "on-the-fly" described in this chapter solve these difficulties. The key idea underlying training on-the-fly is that a model of the process generating the live training data can be used to augment the training set with additional realistic patterns. By modeling both the imaging process and the steering behavior of the human driver, training on-the-fly generates patterns with sufficient variety to allow artificial neural networks to learn a robust representation of individual driving domains. The resulting networks are capable of driving accurately in a wide range of situations.

Acknowledgements

The principle support for the Navlab has come from DARPA, under contracts DACA76-85-C-0019, DACA76-85-C-0003 and DACA76-85-C-0002. This research was also funded in part by a grant from Fujitsu Corporation.

LEARNING MULTIPLE GOAL BEHAVIOR VIA TASK DECOMPOSITION AND DYNAMIC POLICY MERGING

Steven Whitehead
Jonas Karlsson
Josh Tenenberg

Department of Computer Science,
University of Rochester, Rochester New York 14627

ABSTRACT

An ability to coordinate the pursuit of multiple, time-varying goals is important to an intelligent robot. In this chapter we consider the application of reinforcement learning to a simple class of *dynamic, multi-goal tasks*. Not surprisingly, we find that the most straightforward, monolithic approach scales poorly, since the size of the state space is exponential in the number of goals. As an alternative, we propose a simple modular architecture which distributes the learning and control task amongst a set of separate control modules, one for each goal that the agent might encounter. Learning is facilitated since each module learns the optimal policy associated with its goal without regard for other current goals. This greatly simplifies the state representation and speeds learning time compared to a single monolithic controller. When the robot is faced with a single goal, the module associated with that goal is used to determine the overall control policy. When the robot is faced with multiple goals, information from each associated module is merged to determine the policy for the combined task. In general, these merged strategies yield good but suboptimal performance. Thus, the architecture trades poor initial performance, slow learning, and an optimal asymptotic policy in favor of good initial performance, fast learning, and a slightly sub-optimal asymptotic policy. We consider several merging strategies, from simple ones that compare and combine modular information about the current state only, to more sophisticated strategies that use lookahead search to construct more accurate utility estimates.

1   INTRODUCTION

Reinforcement learning has recently been receiving increased attention among researchers interested in developing autonomous intelligent systems. As a method for robot learning, it has a number of appealing properties; not the least of which are that it is easy to use and extensible; it requires little or no *a priori* task knowledge; and it leads to systems that are both highly reactive and adaptive to non-stationary environments. While reinforcement learning is almost certain to play a role in future autonomous systems, the prominence of that role will be largely determined by the extent to which it can be scaled to larger and more complex robot learning tasks.

This chapter considers the application of reinforcement learning to *multiple goal* control tasks. Defined formally in section 2, *multiple goal* tasks, intuitively, correspond to tasks that require coordination of behavior over time in order to accomplish a series of time-varying goals. For instance, an autonomous robot, in addition to performing a primary function, may at times need to attend to certain maintenance goals (e.g., refueling and repair). Or alternatively, a robot may have several goals whose priorities vary with time (e.g., doing dishes, laundry, dusting, vacuuming, changing the cat litter, etc.). Animal behavior is another example. An animal must sustain itself with food and water; but must also avoid predators, locate a mate, nest, procreate, and tend its young.

While it is relatively easy to formulate multiple goal tasks for reinforcement learning, it is less straightforward to learn them efficiently. In particular, the most straightforward approach, which uses a single monolithic state space, suffers from slow learning, due to state space growth that is exponential in the number of possible goals. To avoid this *curse of dimensionality* [Bellman 1957], we propose a modular architecture wherein the overall task is decomposed into its constituent pieces. Following this approach, each individual goal is associated with a control module, whose objective is to learn only to achieve that goal. Decomposing the task in this way, facilitates the learning of individual goals, since information about goal priorities (which is relevant only to coordinating multiple goal activity) is eliminated from each module's internal representation. Instead of learning over a single, monolithic state space, whose size is exponential in the number of goals, the modular architecture learns over a linear number of constant sized state spaces.

When the robot is faced with a single *active*[1] goal, the module associated with that goal is used to determine the overall control policy. When faced with multiple active goals, information from each associated module is merged to determine the policy for the combined task. Since policies for individual goals can be learned much more quickly than a monolithic, global policy, enhanced multi-goal performance can be attained if good policy merging strategies can be found.

Several policy merging strategies are considered. One particularly simple strategy is to select the action associated with the highest utility module.[2] This greedy strategy is easily seen to be non-optimal; however, it yields significantly better learning rates than the monolithic architecture. More sophisticated merging strategies that involve lookahead search are also considered. These techniques are related to recent work in plan merging [Yang 1992, Foulser *et al.* 1990]. In general, using these methods to derive the optimal global policy is NP-hard. However, a class of heuristic branch-and-bound algorithms are developed that exhibit near-optimal performance on a wide range of tasks. Finally, a hybrid architecture that incorporates both modular and monolithic policies is considered. In this architecture, a modular component contributes to good initial performance, while a monolithic component is used to optimize performance in the long run.

Overall, our results indicate that, even though monolithic architectures alone do not scale well to multiple goal tasks, modular architectures that decompose the task by goal are effective, especially when combined with search-based policy merging mechanisms. Moreover, when compared in simulation studies to the hybrid architecture, which was expected to yield better performance in the long run, we found the modular architecture to be surprisingly robust.

---

[1] Here, when we speak of a goal being *active*, we refer to a simplified case in which goals have binary-valued priorities: 0-denoting off (or no priority); 1-denoting on (or high priority). In general, goals may have continuous valued priorities.

[2] In this case, a module's utility is determined by the priority of the goal it represents and by the utility of the current state with respect to that goal.

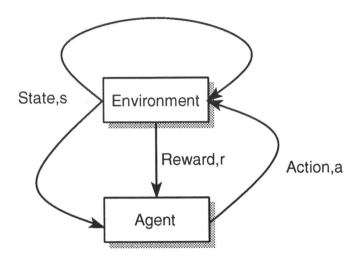

Figure 1.   The basic model of robot-environment interaction.

The remainder of the chapter is organized as follows. Section 2 reviews the basics of reinforcement learning and illustrates them with a simple single goal navigation task. Section 3 discusses multiple goal tasks and considers the learning complexity of the monolithic approach. In Section 4, a modular architecture is presented and shown to significantly outperform the monolithic approach, even when simple strategies are used for policy merging. Section 5 considers search-based merging strategies. A hybrid architecture is discussed in Section 6, and conclusions are drawn in Section 7.

2   BASICS OF REINFORCEMENT LEARNING

Before getting into the details of multiple goal tasks, task decomposition, and merging, it is important to establish the basic formalisms and methods we intend to use in the remainder of the chapter. To that end, we now turn to a brief review of Markov decision processes and Q-learning. For more thorough treatments of each, see [Bertsekas 1987] and [Watkins 1989], respectively. For a review of reinforcement learning in general see [Barto *et al.* 1991].

2.1 Modeling robot-environment interaction

We will assume that the robot-environment interaction can be modeled as a Markov decision process. In a Markov decision process (MDP), the robot and the environment are modeled by two synchronized finite state automatons interacting in a discrete time cyclical process (Figure 1). At each point in time, the following series of events occur.

1. The robot senses the current state of the environment.

2. Based on the current state, the robot chooses an action to execute and communicates it to the environment.

3. Based on the action issued by the robot and its current state, the environment makes a transition to a new state and generates a reward.

4. The reward is passed back to the robot.

Let S denote the set of possible environmental states, and let A denote the set of possible actions. We will assume that both S and A are discrete and finite.

The dynamics of state transitions are modeled by a *transition function*, T, which maps state-action pairs into next states ($T : S \times A \rightarrow S$). In general, the transition function may be probabilistic. If we let x_t denote the state at time t, let a_t denote the action selected at t, and let X_{t+1} be the random variable denoting the state at time $t + 1$, then $X_{t+1} = T(x_t, a_t)$. T is usually specified in terms of a set of *transition probabilities*, $P_{x,y}(a)$, where

$$P_{x,y}(a) = Prob(T(x, a) = y). \qquad (1)$$

Rewards generated by the environment are determined by a *reward function*, R, which maps state into scalar rewards ($R : S \rightarrow \Re$). In general, the reward function may also be probabilistic. If we let R_t be the random variable denoting the reward received at time t, then $R_t = R(x_t)$. For simplicity, we shall hereafter assume that rewards are deterministic, and depend only upon the current state.

Notice that in a MDP, the effects of actions (in terms of the next state and immediate reward received) only depend upon the current state. Process models of this type are said to be memoryless and satisfy the *Markov Property*.[3]

Also notice that the robot has a degree of control over the temporal evolution of the process since it chooses an action at each time step. We shall assume that in selecting control actions the robot's decision is based solely upon the value of the current state.[4] Under these circumstances the robot's behavior can be specified by a control *policy*, which describes the action to execute given the current state. Formally, a policy f is a function from states to actions ($f : S \rightarrow A$), where $f(x)$ denotes the action to be performed in state x.

In reinforcement learning, the robot's objective is to learn a control policy that maximizes some measure of the total reward accumulated over time. In principle, any number of reward measures can be used, however, the most prevalent measure is one based on a discounted sum of the reward received over time. This sum is called the *return* and is defined for time t as

$$return(t) = \sum_{n=0}^{\infty} \gamma^n r_{t+n} \tag{2}$$

where γ, called the temporal discount factor, is a constant between 0 and 1, and r_{t+n} is the reward received at time $t+n$. Because the process

---

[3] Recently, several researchers have become interested in applying reinforcement learning to tasks that cannot easily be modeled using MDPs. These non-Markov tasks typically arise when the robot is unable to properly sense its environment (i.e., cannot sense the state of the environment directly). These tasks introduce difficulties that we would like to avoid here, so we will focus on MDPs only. However, for further information the interested reader may wish to consult [Whitehead 1991b, Tan 1991, Chapman and Kaelbling 1991, Chrisman 1992, Schmidhuber 1990].

[4] A fundamental property of MDPs is that knowledge of the current state is precisely the information needed to perform optimally (with respect to maximizing reward). Therefore, even though it may be possible to devise decision strategies that depend upon additional (stored) information, they cannot possibly outperform the best decision strategies that depend only upon the value of the current state!

may be stochastic, the robot's objective is to find a policy the maximizes the *expected return*.

For a fixed policy f, define $V_f(x)$, the *value function* for f, to be the expected return, given that the process begins in state x and follows policy f thereafter. The robot's objective is to find a policy, f^*, that for every state maximizes the value function. That is, find f^*, such that

$$V_{f^*}(x) = \max_f V_f(x) \qquad\qquad \forall x \in S. \qquad (3)$$

An important property of MDPs is that f^* is well defined and guaranteed to exist. In particular, the *Optimality Theorem* from dynamic programming [Bellman 1957] guarantees that for a discrete time, discrete state Markov decision problem there always exists a deterministic policy that is optimal. Furthermore, a policy f is optimal if and only if it satisfies the following relation:

$$Q_f(x, f(x)) = \max_{a \in A}(Q_f(x, a)) \qquad\qquad \forall x \in S \qquad (4)$$

where $Q_f(x, a)$, called the *action-value* for state-action pair (x, a), is defined as the return the robot expects to receive given that it starts in state x, applies action a next, and then follows policy f thereafter [Bellman 1957, Bertsekas 1987]. Intuitively, Equation 4 says that a policy is optimal if and only if in each state, x, the policy specifies the action that maximizes x's action-value. That is,

$$f^* = a \quad \text{such that} \quad Q_{f^*}(x, a) = \max_{b \in A}[Q_{f^*}(x, b)] \qquad\qquad \forall x \in S. \qquad (5)$$

For a given MDP, the set of action-values for which Equation 4 holds is unique. These values are said to define the optimal action-value function Q^* for the MDP.

If an MDP is completely specified *a priori* (including the transition probabilities and reward distributions) then an optimal policy can be computed directly using techniques from dynamic programming [Bellman 1957, Ross 1983, Bertsekas 1987]. Because we are interested in robot learning, we shall assume that only the state space S and set of possible actions A are known *a priori* and that the statistics governing T and R are *unknown*. Under these circumstances the robot cannot compute the optimal policy directly, but must explore its environment and learn an optimal policy by trial-and-error.

$Q \leftarrow$ a set of initial values for the action-value function (e.g., all zeroes)
For each $x \in S$: $f(x) \leftarrow a$ such that $Q(x, a) = \max_{b \in \mathbf{A}} Q(x, b)$
Repeat forever:
    1) $x \leftarrow$ the current state
    2) Select an action a to execute that is usually consistent with f
       but occasionally an alternate. For example, one might choose
       to follow f with probability p and choose a random action
       otherwise.

    3) Execute action a, and let y be the next state and r be the
       reward received.
    4) Update $Q(x, a)$, the action-value estimate for the state-action
       pair (x, a):
$$Q(x, a) \leftarrow (1 - \alpha)Q(x, a) + \alpha[r + \gamma U(y)]$$
       where $U(y) = Q(y, f(y))$.
    5) Update the policy f:
$$f(x) \leftarrow a \text{ such that } Q(x, a) = \max_{b \in \mathbf{A}} Q(x, b)$$

Figure 2. A simple version of the 1-step Q-learning algorithm.

2.2   Q-learning

In our work we have focussed on a single reinforcement learning algorithm called Q-learning [Watkins 1989]. Other, similar reinforcement learning algorithms could have been used as well with much the same results; however, we shall focus on just this one. For alternatives see for instance [Barto *et al.* 1983, Holland 1986, Williams 1987, Schmidhuber 1990].

In Q-learning the robot estimates the optimal action-value function directly, and then uses it to derive a control policy using the local greedy strategy mandated by Equation 5. A simple version of a Q-learning algorithm is shown in Figure 2. The first step of the algorithm is to initialize the robot's action-value function, Q. Q is the robot's estimate of the optimal action-value function. If some prior knowledge about the task is available, that information may be encoded in the initial values, otherwise the initial values can be arbitrary (e.g., uniformly zero). Next the robot's initial control policy, f, is established. This is done by assigning to $f(x)$ the action that locally maximizes the action-value. That is,

$$f(x) \leftarrow a \quad \text{such that} \quad Q(x, a) = \max_{b \in A} Q(x, b), \tag{6}$$

where ties are broken arbitrarily. The robot then enters a cycle of acting and policy updating. First, the robot senses the current state, x. It then selects an action a to perform next. Most of the time, this action will be the action specified by the robot's policy $f(x)$, but occasionally the robot will choose a random action.[5] The robot performs the selected action and notes the immediate reward r and the resulting state y. The action-value estimate for the state action pair (x, a) is then updated. In particular, an estimate for $Q^*(x, a)$ is obtained by combining the immediate reward r with a utility estimate for the next state, $U(y) = \max_{b \in A}[Q(y, b)]$. The sum

$$r + \gamma U(y), \tag{7}$$

called a 1-step corrected estimator, is an unbiased estimator for $Q^*(x, a)$ when $Q = Q^*$, since, by definition

$$Q^*(x, a) = E[R(x, a) + \gamma V^*(T(x, a))], \tag{8}$$

where $V^*(x) = \max_{a \in A} Q^*(x, a)$. The 1-step estimate is combined with the old estimate for $Q(x, a)$ using a weighted sum:

$$Q(x, a) \leftarrow (1 - \alpha)Q(x, a) + \alpha[r + \gamma U(y)], \tag{9}$$

where α is the learning rate. Finally, the robot's control policy is updated using Equation 5, and the cycle repeats.

2.3   An Example

As a simple example consider the grid world problem shown in Figure 3. In this example, the robot is free to roam about a bounded 2-dimensional

---

[5] Occasionally choosing an action at random is a particularly simple mechanism for exploring the environment. Exploration is necessary to guarantee that the robot will eventually learn an optimal policy. For examples of more sophisticated exploration strategies see [Kaelbling 1990, Thrun 1992, Sutton 1990].

grid. It can move in one of four principle directions, left, right, up, or down. The robot has "wall" sensors that tell it the distance to the upper and left walls. The effects of actions are non-deterministic, but not completely random. With high probability an action will have its "intended" effect (e.g., the left action will cause the robot to move one position to the left, a right action, one position to the right, etc); however, occasionally an action may result in the robot moving in an unexpected direction. In this case, the motion is in a neighboring direction (e.g., unexpected results of a "right" action result in either motion "up" or "down", etc.) The robot's objective is to navigate to the labeled goal position. To entice the robot to the goal position, a small positive reward is generated each time it enters the goal state. In other states, the robot receives no reward. To facilitate complete exploration of the state space, each time the robot enters the goal state, it is teleported to a new random location in the grid. In this way, the temporal evolution of the process resembles a sequence of repeated trials.

The state space for this task is determined by the possible values for the location sensors. In this case, the grid is 20 × 20, so there are a total of 400 distinct states. With a choice of four possible actions per state, the robot must estimate a total 1600 action-values.[6] If the underlying structure of the environment and the position of the reward are initially unknown, then it is difficult to accurately estimate the optimal action-value function. Under these circumstances a particularly simple approach is to initialize all action-values to zero. In this case, the robot's initial performance will be random (assuming ties are broken by choosing randomly), and learning first occurs when the robot first encounters the goal state. At that point, the action-value for the state-action pair that immediately proceeded the goal state is increased. On subsequent trials, the action-values of other state-action pairs are incrementally increased as they are found to lead to either reward states or states of high utility. In this way, utility information is "backed up" until an accurate

---

[6] In our experiments we have used tables to implement the action-value function. The advantage of using tables is that they are conceptually simple and easy to use. The disadvantages are that they can require substantial amounts of memory and they provide no mechanism for generalizing (or interpolating) between neighboring state-action pairs. Lack of generalization can result in slow learning. In general, any number of function approximating techniques can be used to implement the Q-function (e.g., neural networks, or CMACS); however, the primary issues addressed in this chapter are largely independent of the particular implementation details, so, for clarity, we have chosen to stick to the simplest possible realization.

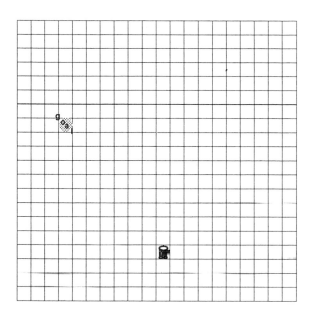

Figure 3. A simple reinforcement learning task. The robot receives a reward when it finds the goal state.

estimate of the optimal action-value function is obtained and an optimal control policy is learned.

The time needed to learn an optimal policy depends on many factors. One important factor is the size of the state space. In particular, when reward is sparse (e.g., when there is only a single goal state), the expected learning time for the robot outlined above is exponential in the size of the state space [Whitehead 1991a]. Much of this learning time is spent during the first few trials when the robot explores the environment more or less randomly. There are a number of methods that can be employed to speed up learning in general. One technique is to adapt more sophisticated strategies for exploring the environment [Kaelbling 1990, Thrun 1992, Sutton 1990]. For example, [Thrun 1992] shows that in certain deterministic environments, the learning time can be reduced to a small polynomial factor by having the agent incrementally learn a "map" of the environment and then use it to plan experiments into previously unexplored regions. Another technique is to employ more sophisticated updating procedures. For instance, [Lin 1991a] has demonstrated an "experience replay" algorithm that in effect rehearses successful trials until the maximum benefit of the experience has been achieved. Similarly, incremental planning methods perform a similar function [Sutton 1990, Whitehead and Ballard 1989]. The learning rate can also be improved by enriching the learning environment with additional sources of feedback and experience. For instance, [Whitehead and Ballard 1991] and [Whitehead 1991a] explore the benefits of "Learning-with-an-External-Critic" (LEC) and "Learning-By-Watching" (LBW).[7] In LEC, an external supervisor monitors the agent's performance and occasionally provides feedback (rewards and penalties) indicating the appropriateness of the agent's immediate behavior. This immediate feedback effectively eliminates the temporal credit assignment problem [Sutton 1984] and significantly reduces the requisite learning time. In LBW, the agent is assumed to be capable of observing the behavior of one or more additional agents that are engaged in the same task. By observing the skilled performance of an external role model, a naive agent gains valuable experience, avoids unproductive random exploration, and learns much faster. While each of these methods are effective within the context of a single goal task, they all suffer from the curse of dimensionality (i.e., exponential growth in learning time) when we scale to multi-goal tasks.

3 MULTIPLE GOAL TASKS

The example task outlined in Figure 3, though useful for illustrating the principles of reinforcement learning, is extremely simple and not particularly representative of the control problems facing autonomous robots. One degree of complexity that is conspicuously missing in the above example is the notion that intelligent behavior involves the coordination of multiple activities. In the example, the robot's only goal is to navigate to the goal state as quickly as possible. We are interested in studying agents that coordinate behavior in order to accomplish a series of tasks that change over time. For example, we can imagine an animal having such goals as getting food, getting water, procreating, tending to young, avoiding predators, and so on. Such goals are more dynamic and cyclic than a single fixed goal. Drives, purposes, and goals come and go with

---

[7] Also see [Clouse and Utgoff 1992] for another example of LEC.

time, as resources are consumed and renewed, as circadian cycles repeat, as opportunities and interruptions come and go. Individual goals may come and go repeatedly over time and may arise in arbitrary combinations with other goals. Accordingly, a robot's behavior should be influenced by the set of goals that are currently active.

The example task in Figure 3 can be elaborated into a more interesting multiple goal task as follows. First let us introduce additional goals. In particular, instead of having just a single goal, suppose the robot has n goals, $\Gamma^1, \Gamma^2, \ldots, \Gamma^n$. Next, instead of having each goal be satisfied by a single state, associate with each goal Γ^i a set of satisfying states $G^i \subset S$. Finally, associate with each goal Γ^i a reward function R^i such that

$$R^i(x) = \begin{cases} c^i & \text{if } x \in G^i \\ 0 & \text{otherwise.} \end{cases} \tag{10}$$

where c^i is a scalar constant reward associated with each goal Γ^i.

This extension only goes part way towards our desired objective. At this point we have multiple goals; but the goals are static. To obtain time varying goals we introduce the notion of goal activations. That is, we assume that each goal has associated with it a time dependent activation, which we denote by $g^i{}_t$. For simplicity we shall assume that activations are binary valued processes, where $g^i{}_t = 1$ indicates that, at time t, Γ^i is active, and $g^i{}_t = 0$ indicates Γ^i is inactive. At each time point, then, we view each g^i as a bit, and the vector of all such bits we call the activation vector:

$$\bar{g} = g^1 \cdot g^2 \cdot g^3 \cdots g^n. \tag{11}$$

Intuitively, a goal's activation encodes its current importance. Functionally, activation values are used to modulate the reward received by the robot. In particular, we assume that the robot only receives a reward upon entering a goal state for an active goal. Under this new scheme, the reward function depends on both the current world state and the current activation vector:

$$R(x, \bar{g}) = \sum_{i=1}^{n} R^i(x)g^i. \tag{12}$$

We assume that the dynamics of goal activations are described by the following transition rules: If $g^i_t = 1$, then

$$g^i_{t+1} = \begin{cases} 0 & \text{if } x_t \in G^i \\ 1 & \text{otherwise.} \end{cases} \tag{13}$$

If $g^i_t = 0$, then

$$g^i_{t+1} = \begin{cases} 1 & \text{with probability } p \\ 0 & \text{with probability } 1 - p. \end{cases} \tag{14}$$

Using these transition rules, a goal, once activated, persists until it is achieved. Inactive goals on the other hand, spontaneously become active according to a Bernoulli process. Multiple goal tasks of this form seem to more closely model the dynamics of tasks facing autonomous systems since different goals come and go over time and as the robot attends to its various needs as they arise. Notice that in a multiple goal task there is no need to artificially partition the robot's behavior into trials; nor is it necessary to introduce operations that magically teleport the robot to new points in the state space once a goal is reached. In a multiple goal task, when the robot accomplishes one goal, that source of reward disappears (at least temporarily) and the robot naturally moves on to perform other activities (and seek alternative sources of reward).[8]

Multiple goal tasks can be modeled as Markov decision processes by extending the state space to include both the state of the external world and state of the activation vector. Formally, a multiple goal task is defined by the MDP (S_m, A_m, T_m, R_m), where

> S_m is a composite state space that encodes both external world state and goal activations. $S_m = S \times \bar{G}$, where S is the set of possible world states and $\bar{G}$ is the set of possible goal activation vectors. For an n-goal task, $\bar{G} = \{0, 1\}^n$.

> A_m is the set of actions available to the robot.

---

[8] It is certainly possible to extend our definition further by introducing continuous valued goal activations and activation rates that are goal specific. Similarly, it is straightforward to devise transition rules for modeling periodic processes. Our assumption that goals are statistically independent is also a simplification. The goal activation rules given above serve as a useful jumping off point (in that they provide for the notion of repeated, semi-unpredictable, multiply-combined goal activations). However, future research is needed to develop and understand the interactions associated with more sophisticated models.

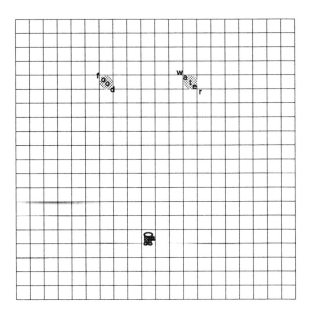

Figure 4. A simple multiple goal task.

T_m is the composite transition function on $S \times \bar{G} \times A_m$ into $S \times \bar{G}$. Specifically,

$$(x_{t+1}, \bar{g}_{t+1}) = T_m(x_t, \bar{g}_t, a_t) \tag{15}$$

where

$$x_{t+1} = T'(x_t, a_t), \tag{16}$$

$$\bar{g}_{t+1} = T_A(\bar{g}_t, x_t), \tag{17}$$

T is the underlying transition function for external world states, and T_A is the transition function for goal activations which encodes the rules given in Equations 13 and 14.

R_m is the composite reward function given by Equation 12.

We shall use f_m^* and Q_m^*, respectively, to denote the optimal policy and optimal action-value function for a given multi-goal task.

To illustrate, consider the example shown in Figure 4. The robot is in the bottom part of the grid, and the other marked cells indicate the locations

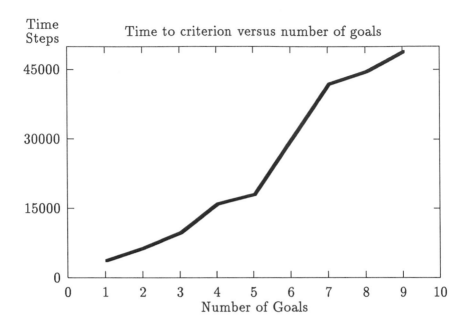

Figure 5. The time needed by a monolithic Q-learning system to reach the criterion of "relatively good performance" as a function of the number of goals in a task similar to the one shown in Figure 4. See text for the definition of "relatively good performance".

where different goal objects (e.g., food and water) can be found. The set of actions are, as before, the movements into adjacent cells. The states, however, encode not only position, but also the set of goals that are currently active (i.e., whether the robot needs to get food, water, or both). The robot's need for a particular object arises according to a Bernoulli process, and once the object is obtained, the robot receives a reward, and the associated goal activation is set to 0. The robot then turns its efforts toward achieving its remaining active goals.

Notice that our definition of multiple goal task is somewhat different than others found in the literature (e.g., [Mahadevan and Connell 1991, Singh 1992]). In [Singh 1992] a *composite* task is defined by sequentially concatenating multiple *elemental* tasks. In a composite task, rewards are generated only when the system achieves a subtask (goal) in a prescribed order. In effect, in a composite task only one goal is active at

a time. Although related, Singh's objective is substantially different
from ours. Singh's objective is to develop systems that can efficiently
learn long, complex sequences of behavior by first learning simpler elemental sequences, and then learning to compose them. Our objective
is to study system that coordinate behavior in order to pursue multiple, independent goals in parallel. In our model, we assume active goals
can be achieved in any order that is convenient. Similarly, the work
described in [Mahadevan and Connell 1991] is more closely related to
that of Singh, in that Mahadevan and Connell describe a system that
decomposes a single complex activity into a series of subactivities, each
of which is learned by an individual module. This sequential decomposition substantially improves the learnability of such complex sequential
tasks. However, it does not address the issue of coordinating behavior
between parallel competing goals.

3.1   Learning Multiple Goal Tasks

Because multiple goal tasks (as we have defined them), are MDPs, we
know that Q-learning can be used to construct robots that are guaranteed to eventually learn an optimal control policy. However, the straightforward application of Q-learning, in a *monolithic* controller, leads to
prohibitively slow learning. In particular, scaling from a single goal
task to a n-goal task leads to an exponential growth in the state space.
Instead of having to learn action values for $|S| \cdot |A|$ state-action pairs,
$|S| \cdot 2^n \cdot |A|$ action-values must be learned. In general, if goal activations
are allowed to take on a range of values, say m (where value encodes
priority) the size of the state space scales as m^n. Without any means of
generalizing (or interpolating) action-value estimates across state-action
pairs, the learning time can be expected to scale exponentially in the
number of goals.[9] Indeed, this poor performance prediction is verified
in Figure 5, which shows the time needed by a monolithic Q-learning
robot to reach a criterion of reasonably good performance on a series of
multiple goal tasks in the 2-D grid world.[10]

---

[9] Even if techniques are used that provide for generalization (or interpolation), such as
CMACs or neural networks, they are unlikely to be effective since a small change in the
activation vector (e.g., the activation/deactivation of a single goal) can profoundly affect the
action-value.

[10] Since the optimal policy is difficult to compute, and requires a very long time to learn
absolutely, we have adopted a performance criterion based on "reasonably good performance."

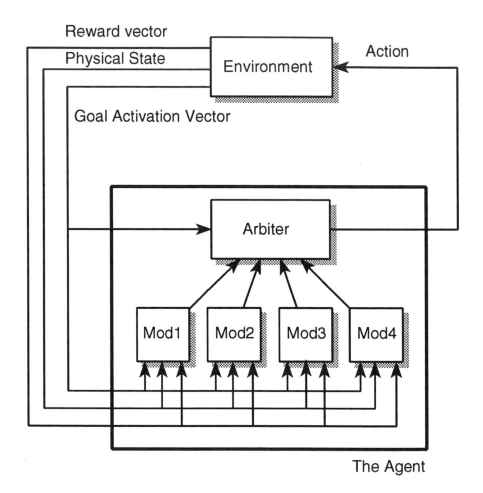

Figure 6.   The modular architecture. Each module is responsible for learning to achieve a single goal. The arbiter is responsible for merging information from the individual modules in order to derive the single action performed by the robot at each time step.

4   A DECOMPOSITION APPROACH

The trouble with the monolithic approach is that experience gained in solving a goal under one set of goal activations is not easily transferred to solving that same goal under another similar goal activation set. For example, in the task shown in Figure 4, the action-values updated when the active goals are get food and get water are distinct from the action-values updated when just the goal get food is active. Consider the case where the robot is very close to the food, and has only the goal of getting food. In this case, the robot should clearly head directly for the food. Now consider the case where the robot is located at the same point in the environment, but has active goals for getting both food and water. In this case, the robot should also head directly toward the nearby food, satisfy that goal first, then move toward the water. Unfortunately, the monolithic architecture does not provide any mechanisms for transferring control knowledge from the single goal case to the dual goal case. On the other hand, transfer is not always warranted. For example, if instead of being near the food, the robot was closer to the water, then when both goals are active, the agent should pursue the water first. In this case, it is inappropriate to apply the actions used when just the goal for getting food was active. In general, the action-values for situations with identical world states but different goal activations are bound to be different since different combinations of goals require different optimal solution strategies. Nevertheless it seems intuitively clear that a multi-purpose robot should be capable of encapsulating knowledge in a goal-specific way and transferring experience/knowledge about attaining a goal in one set of goal activations to another.

4.1   The modular architecture

The modular architecture shown in Figure 6 provides this capability. In the modular architecture a set of independent fixed sized Q-learning modules are used. Each module is itself an adaptive controller and is responsible for learning to achieve a single goal. The state space of

---

In our experiments, we have defined reasonably good performance to be achieved when the robot can consistently obtain reward at a rate equivalent to 90% of that obtained by a robot using an "optimal-greedy strategy" — that is a robot that at each point in time takes an action that is optimal with respect to achieving the nearest goal. This strategy is suboptimal with respect to the overall multiple goal task, but yields reasonably good performance.

each module encodes only information about the external world (goal activation information is not represented). The set of states, actions, and the transition function for each module are taken to be the same, so that modules differ only in their reward function.[11]

Modules learn to achieve their respective goals in parallel. That is, on each step, each *active* module updates its action-value function according to the update rules for Q-learning.[12] However, instead of assuming that a single composite reward value is generated at each time step, we assume that a reward vector is generated, where each component in the vector corresponds to the reward value of a single goal. Thus, when a goal is achieved and a reward is generated, that goal's reward is routed directly (and only) to the module responsible for learning that goal. Also, when a module updates its action-values it uses goal specific utility estimates derived from its local action-value function. In this way each module is specialized to a specific goal. To facilitate the propagation of utility information within a module, each *active* module is updated at each time step, regardless of the reward it receives and regardless of whether or not it was responsible for generating the current action.

Since individual modules may specify conflicting actions, an arbitration module is used to mediate global control. This arbiter receives as input both the world state and the global activation vector. At each time step, it generates the single action performed by the robot. There are a variety of algorithms that can be used to implement the arbiter. We have focussed on algorithms that construct estimates for the global action-value function, Q_m^*, by merging action-values from the individual modules. A range of merging strategies can be defined by trading off solution quality against computational complexity. The most elementary strategies combine and compare the modular action-values for the current state only, whereas more sophisticated strategies use predictive models and lookahead search to construct more accurate estimates.[13]

---

[11] Actually, all of our algorithms assume only that there is a common set of actions defined across all modules, allowing for differences in state spaces and transition functions between modules. This might occur, for example, if there are input bits encoding features of the domain that are only relevant to particular goals.

[12] We refer to a module as active, if its associated goal is active. Only active modules are updated, since rewards are only generated when goals are active, and since we only want a module to know how to achieve its goal when the goal is active.

[13] While the structure of the architecture shown in Figure 6 resembles those of [Singh 1992] and [Mahadevan and Connell 1991], they are more different than they are similar. For

4.2   Simple Merging Strategies

We will consider a merging strategy to be *simple* if

1. only the next action to be executed is computed, as opposed to extended action sequences, and

2. the time to compute the next action is no larger than a small polynomial in the number of goals.

The first condition obviates the need for a detailed predictive model, while the latter condition constrains the kinds of computation that are considered.

One simple merging strategy that we have explored is the *nearest neighbor* strategy, whose global policy we denote by f_{nn}. When using nearest neighbor, the arbiter chooses the action among all active modules that has the maximal action value. That is,

$$f_{nn}(x, \bar{g}) = \arg\max_{a \in A} Q_{nn}(x, \bar{g}, a), \tag{18}$$

where

$$Q_{nn}(x, \bar{g}, a) = \max_{i \in [1...n]} [Q^i(x, a) \cdot g^i]. \tag{19}$$

and where Q^i denotes the action-value function maintained by the ith module. Returning to the situation illustrated in Figure 4, the robot must choose between going to water and going to food. Assume each module has learned an accurate action value function for its goal, and that the rewards for food and water are identical. If the food is closest, the policy action associated with the *get-food* module will have the highest action-value, and hence this action will be chosen by f_{nn}. Once food is obtained, the robot will move toward the water, since this is the only remaining active goal. Because each module's policy corresponds to performing a gradient ascent in the utility space, an agent using the

---

instance, in [Singh 1992] a single submodule, which is selected by a gating module, is in control at each point in time. In [Mahadevan and Connell 1991] the task is partitioned such that only one module is active at a time. In our architecture, the arbiter is always in control of the agent, and the actions it chooses may or may not correspond to the actions prescribed by active submodules. Also our architecture assumes that multiple goals (and modules) can be active at a time, and that a reward can be distinguished (and routed) according to the goal that generated it.

nearest neighbor strategy will tend to complete its current goal, using the policy from a single module, before switching to another goal and a different module.

Unfortunately nearest neighbor may perform poorly because it considers goals individually. In the case where a cluster of distant goals has a higher cumulative value than a single nearby goal, nearest neighbor will prefer the nearby goal to the cluster, even when the optimal strategy is to pursue the cluster.

An alternative strategy, which we call *greatest mass*, accounts for goal clustering. Using this strategy, the next action chosen by the arbiter is the one that maximizes the sum of the expected values across all modules. That is,

$$f_{gm}(x, \bar{g}) = \arg\max_{a \in A} \left[Q_{gm}(x, \bar{g}, a) \right], \tag{20}$$

where

$$Q_{gm}(x, \bar{g}, a) = \sum_{i=1}^{n} \left(Q^i(x, a) \cdot g^i \right). \tag{21}$$

The intuition behind the greatest mass strategy is for the robot to move toward regions of the state space with the most reward. Methods analogous to it have been used in several path planning systems. In [Arbib and House 1987], a frog's trajectory planning is modeled by letting obstacles and targets (a fly) rate different directions of travel, and chosing the direction with the highest combined rating. Similarly, [Khatib 1986] assigns positive and negative potentials to obstacles and goals, which combine to form a potential field that the robot traverses using a gradient descent.

With greatest mass, the robot may forgo an immediate reward from a nearby target in favor of multiple rewards from distal sources. Generally, the trade off depends upon the number, magnitude, and distance to the rewards and upon the strength of the temporal discount. The less future reward is discounted, the more likely closer goals will be abandoned in favor of a distant cluster. However, for heavily discounted rewards ($\gamma \ll 1$), the greatest mass strategy, like nearest neighbor, will pursue nearby goals first.

These simple strategies have the following desirable characteristics:

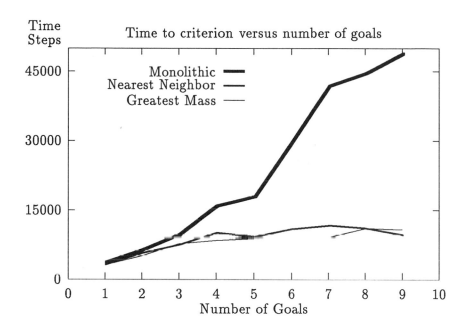

Figure 7. The time needed to reach the "reasonable-performance" criterion as a function of the number of goals in a multi-goal task for two modular systems and a monolithic system. The learning time required by the modular systems is largely independent of the number of goals.

1. In each state, arbitration takes time linear in the number of goals.

2. Since only local action values are used for arbitration, simple strategies are appropriate in domains where prediction is difficult (e.g., non-deterministic domains).

3. Under conditions of high temporal discount ($\gamma \ll 1$), both strategies closely approximate the optimal policy (assuming the individual goal modules have accurate action-value functions).

4.3   Performance evaluation

To evaluate the potential performance improvement available from the modular architecture we tested it on a series of tasks on a 20x20 grid. In particular we ran experiments where the number of goals ranged from 1 to 9 in increments of 2, and measured the time needed to reach our "reasonable performance" criterion. Figure 7 shows performance curves for two modular systems, one using the nearest neighbor merging strategy and the other using the greatest mass strategy. Also shown is the performance of the comparable monolithic system. The figure shows that, unlike the monolithic system, the time needed to reach criterion for the modular system is largely independent of the number of goals. An alternative view can be obtained from Figure 8 which plots the average cumulative reward received over time for each of the three systems on a 9-goal task. This figure clearly shows the enhanced performance of the modular systems.

In general, the modular architecture trades off optimal performance in the limit, for faster learning up front. This follows since, even if each goal module learns to achieve its goal optimally, global performance is limited by suboptimal arbitration. On the other hand, decomposing the system into goal-achieving modules, substantially reduces the learning time needed to reliably achieve each goal. Since even simple merging strategies yield near-optimal performance (in a wide range of cases) and since for many applications satisficing behavior is sufficient, the trade off seems justified.

5   SEARCH-BASED MERGING

The simple merging strategies described in the previous section only approximate the action-value function for the global task. This follows since the action-values of a given module only encode the expected return given that the robot perform some action, and then follow the policy that is optimal with respect to the module's goal. Whereas, the global action-value function must encode the expected return given that the robot perform some action, and thereafter follow the policy that is *optimal with respect to all the goals combined*. If we assume that the action-value functions for the individual modules are accurate, then, in general, Q_{nn} is guaranteed to underestimate the global action-value function and Q_{gm} tends to overestimate.

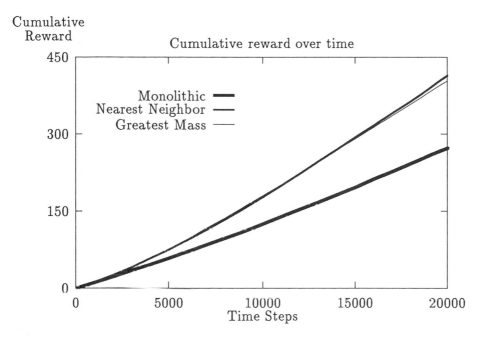

Figure 8. Cumulative reward versus time for two modular systems and a monolithic system.

More accurate global action-value estimates, and hence better global performance, can be achieved by using merging strategies that use lookahead search. To simplify the discussion, assume, for the remainder of this section, 1) that each individual goal module has learned an accurate action-value function for its goal, 2) that the external environment is completely deterministic, and 3) that the robot has at its disposal a predictive model that it can use to perform mental simulations of possible action sequences.[14] Given these assumptions, one way for the arbiter to proceed is to generate a search tree of possible action sequences; compute (via mental simulation on the predictive model) the utility of each path in the tree; and then select for execution the first action in the sequence with the maximal utility.

---

[14] The predictive model can be of any form, but it generally contains information encoded in the global Transition and Reward functions, T_m and R_m, respectively. Note, our assumption that the agent has a predictive model is a strong one. In general the time needed to learn such a model may be prohibitive. For instance see [Sutton 1991, Mahadevan 1992, Pierce 1991a]. Nevertheless, if the agent is continually encountering different combinations of goals, then the cost of learning the model can be traded off against a lifetime of improved global performance.

Clearly mentally searching all possible sequences for a best possible path is intractable since the search space grows exponentially with path length. Therefore heuristic methods must be used to generate a set of plausible sequences to search. Since the purpose of the arbiter is to select the single action to perform next, one useful set of sequences to search (assuming for the moment that they could be generated) would be those sequences that begin by performing some action $a \in A$ and then follow the optimal global policy until all active goals have been achieved. Of course, the global optimal policy, is not known *a priori*, so this set of search paths cannot be generated directly. However, if an estimate of the optimal global policy can be obtained, then the set of sequences that result from replacing the optimal policy with an approximation, can be used as an alternative.

This idea has lead to an arbitration strategy which we call *Lookahead Nearest Neighbor* (LANN). In LANN the arbiter defines, for each action $a \in A$, p_a^{nn} to be the sequence that results from first performing action a in the initial state $(x_0, \bar{g}_0)$, and then following the nearest neighbor policy, f_{nn}, until each active goal has been achieved. That is,

$$p_a^{nn} = \{(x_0, \bar{g}_0), a, (x_1, \hat{\bar{g}}_1), f_{nn}(x_1, \hat{\bar{g}}_1), (x_2, \hat{\bar{g}}_2), f_{nn}(x_2, \hat{\bar{g}}_2),$$
$$...(x_{m-1}, \hat{\bar{g}}_{m-1}), f_{nn}(x_{m-1}, \hat{\bar{g}}_{m-1}), (x_m, \hat{\bar{g}}_m)\}.$$

We define $Q_{\text{LANN}}((x, \bar{g}), a)$ to be the expected return given that the agent, starting in state x follows the sequence p_a^{nn} until all active goals are satisfied and follows the optimal policy thereafter. Also we define $f_{\text{LANN}}(x, \bar{g}) = \arg\max_{a \in A} Q_{\text{LANN}}((x, \bar{g}), a)$. Notice that if f_{nn} closely estimates f_m^*, then Q_{LANN} will better estimate Q^* than Q_{nn}, and f_{LANN} will be a better global policy than f_{nn}. This follows since Q_{LANN} is constructed by projecting along a path that is nearly optimal with respect to the global task and summing the rewards obtained, whereas Q_{nn} is based on the return estimates associated with pursuing only the nearest (or highest utility) goal and ignoring all other goals and their possible interactions. A pseudo-code algorithm for computing Q_{LANN} is shown in Figure 9.

Function Compute-Q-LANN(x_0, $\bar{g}_0$, a_0);
  { Computes the action-value estimate for the
    Lookahead Nearest Neighbor Merging Strategy}

    d ← 0;   {search depth so far}
    currentState ← $T(x_0, a_0)$;
    $\bar{g} \leftarrow \bar{g}_0$;
    totalReturn ← 0;
      for each active goal y^i
        if R^i(currentState) > 0 then
            totalReturn ← totalReturn + R^i(currentState)(γ^d);
            $g^i \leftarrow 0$;   {deactivate this goal}
        endif
      endfor
    while (there are active goals in $\bar{g}$) do
      currentAction ← $\arg_a \max_{a \in A, i \in [1...n]} \left[Q^i(x, a) \cdot g^i \right]$;
      currentState ← T(currentState, currentAction);
      d ← d + 1;
      for each active goal g^i
        if R^i(currentState) > 0 then
            totalReturn ← totalReturn + R^i(currentState)(γ^d);
            $g^i \leftarrow 0$;   {deactivate this goal}
        endif
      endfor
    endwhile
    return totalReturn;

Figure 9. Computing Q_{LANN}.

5.1   An example

An example will help to illustrate the potential benefits of search-based merging strategies in general and the LANN strategy in particular. Consider an extension of the two goal task shown in Figure 4, where the robot is given a reward for completing either of getting food or getting water, but only after returning with the food or water to a designated starting location (the nest). For simplicity, assume the robot is initially positioned at the nest. Assume the reward for bringing the food to the nest is sixty units, and the reward for water is fifty. In addition assume that future reward is moderately discounted (say, $\gamma = 0.95$). The distance from nest to water is 14, from nest to food, 14, and from water to food, 6.

From the initial state, the candidate actions are to go either toward the water or the food, denoted a_w and a_f respectively. The sequence for $p^{nn}_{a_w}$ takes the robot first to the water, back to the nest, onto the food, and finally back to the nest, for an estimated path utility of $(50)(\gamma^{28}) + (60)(\gamma^{56}) = 15.3$. This value is obtained by summing the appropriately discounted rewards received along path generated. Since the path length for receiving the reward for water is 28, this reward is discounted by γ^{28}. Likewise for receiving food, where the path length from the initial state is 56.

The sequence for $p^{nn}_{a_f}$ first arrives at the food, returns to the nest, goes to the water, and then returns to the nest, for an estimated path value of $(60)(\gamma^{28}) + (50)(\gamma^{56}) = 17.1$. Since the utility of the former path is greater than that of the latter, an arbiter using LANN selects a_f for its first action.

Initially f_{LANN} generates the same trajectory as f_{nn}. In particular, it will cause the robot to follow a straight path to the food. However, after reaching the food (but before returning to the nest), the behavior of the LANN strategy diverges from that of nearest neighbor. In particular, once the food has been reached, the robot's candidate actions are to return to the nest or head for the water, denoted a_n and a_w respectively. From this point, the search path generated by first moving toward the nest, $p^{nn}_{a_n}$, leads the robot back to the nest, onto water and then back, and has a estimated path utility of $(60)(\gamma^{14}) + (50)(\gamma^{42}) = 35.1$. The search path generated by first moving toward the water, $p_{a_w}^{nn}$, leads the robot first to the water, then to the nest where it receives both rewards simultaneously. This path has an estimated utility of $(60)(\gamma^{20}) + (50)(\gamma^{20}) = 39.4$. Thus, a robot using the LANN merging strategy will, upon reaching the food location, temporarily defer its goal to get food in order to set up the water goal and eventually receive a larger cumulative payoff in the end.

As with the greatest mass strategy, the willingness of the LANN strategy to defer its current goal and temporarily pursue another depends upon the temporal discount and the relative goal strengths. When $\gamma = 0.90$, the sequence values for $p_{a_n}^{nn}$ and $p_{a_w}^{nn}$ are 14.3 and 13.4, respectively. That is, when future reward is heavily discounted, the robot tends to adopt a greedy strategy. However, when $\gamma = 0.97$, the sequence values are 53.1 and 59.8, respectively. As future reward is less heavily discounted, the robot is increasingly willing to defer more immediate reward in favor of a higher overall future return. As γ approaches 1, deferring immediate reward is still preferred, but the difference in utility between doing so and adopting a greedy strategy approaches zero.

If the temporal discount is held constant and the reward is changed, the behavior of the robot is similarly affected. The smaller the remaining reward, the less likely the robot is to defer its pursuit of a more immediate goal.

One of the appealing properties of the LANN strategy is that, when f_{nn} closely estimates f_m^*, LANN is guaranteed to be at least as good if not better than simple nearest neighbor. This follows since LANN generates a nearest neighbor policy unless it can be shown by looking ahead that pursuing another action choice will eventually lead to a higher return.

5.2 Extending the look ahead strategies

The search-based strategies discussed so far can be varied/extended along a number of dimensions.

1. A shorter depth bound can be placed on the look ahead part of the search. That is, the robot can stop search paths before all active goals have been satisfied, using as its sequence value the discounted reward received so far. One method for choosing a shorter depth bound is to terminate the path when the possible discounted reward that can be received from the remaining goals drops below some small threshold.

2. Use a different heuristic for generating sequences. We have not claimed that using nearest neighbor provides the optimal sequences, but rather, that it appears to be quite good. There are other promising candidates, such as greatest mass.

3. Maintain an estimate of the probability with which the different goals arise. This probability can be used to account for the effects of inactive goals in the sequence value estimates and allow the robot to *anticipate* its future need for an inactive goal. For example, using this strategy, if the robot's need for food frequently arose, it would rarely choose actions that brought it far from a known food source.

4. Replace the temporal discount on reward with explicit action costs. In many circumstances, for instance, where each action uses a certain amount of fuel, it is more intuitive to associate fixed costs with each action rather than simply discount future rewards. We have found that when using fixed cost actions and no temporal discounting the optimal strategy is much more likely to involve deferring a more immediate goal in favor of more distant ones.

5.3   Trading off computation time for solution quality

Assuming an accurate predictive model, the search-based strategy that we have outlined is likely to outperform its analogous simple strategy. However, one must clearly balance the increase in solution quality with the added cost of search based method. These costs primarily involve the need to learn, store, and maintain an explicit predictive model, and the time required to compute the search paths. In addition, the search-based strategy we have discussed assume deterministic environments. In

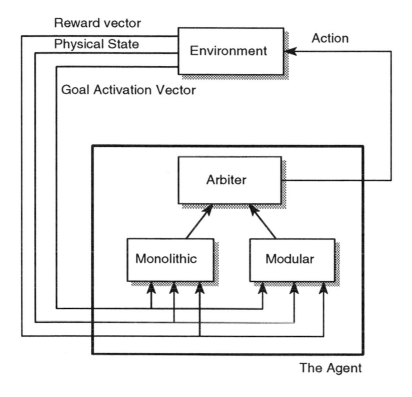

Figure 10. The Hybrid Architecture

less deterministic domains, this strategy (and others like it) will be more costly, and hence, less valuable.

6 A HYBRID ARCHITECTURE

The monolithic approach described in Section 3 leads to optimal control in the limit, but has poor initial performance. Conversely, the modular architecture trades off asymptotic performance in favor of faster learning up front. A hybrid architecture that incorporates components of both may achieve both fast learning and optimal asymptotic performance. A simple hybrid architecture is shown in Figure 10. This architecture has three components: a modular component, a monolithic component, and a modular-monolithic arbiter. The modular component corresponds to a complete modular architecture (see Figure 6), and may use any of the merging strategies described above. The monolithic component is a complete monolithic system (i.e., the monolithic system implements a single, monolithic Q-learning system). The modular-monolithic arbiter (hereafter, the arbiter) is responsible for mediating global control. At each time step, it generates the single action performed by the robot. Early on, we would like the arbiter to select the actions prescribed by the modular component, when it is more likely to yield good performance. However, eventually, we would like the arbiter to rely, more and more, upon the monolithic component.

Unlike the arbiter in the modular component, the modular-monolithic arbiter does not combine utility information from its subcomponents to select an action. Instead, it selects one or the other of the actions proposed by the two sub-components. One selection strategy is simply to choose the action (among the two) with the highest estimated global action-value. That is, associated with the actions proposed by the modular and monolithic components is a global action-value estimate. In the monolithic case, the global estimate is simply the action-value encoded in the monolithic component's action-value function. In the modular component, the global estimate is determined by the particular merging strategy employed. For example, if the nearest-neighbor merging strategy is used, then Q_{nn} can be used. For the greatest mass merging strategy, Q_{gm} can be used. Similar global action-value estimates can be defined for search-based merging strategies.

If the global action-values generated by the modular component are guaranteed to eventually be no greater than the true optimal action-values and if the monolithic component eventually learns the optimal action-value function, then the monolithic system will over time become the dominant component. However, if the modular component's action-values can overestimate the true global action-values, even when the individual modules have accurate Q-functions, then the monolithic component may never dominate and optimal performance may not be achieved. If the Q-functions of the individual modules in the modular component accurately estimate the optimal action-values for their respective goals, then Q_{nn} will either equal or underestimate the global action-values. Conversely, Q_{gm} can easily be shown to overestimate in many cases. Thus, for use in a hybrid architecture, the nearest neighbor strategy may be preferred.

Preliminary experiments indicate that even after 100,000 steps, the modular and hybrid architectures showed roughly the same performance. This suggests that it may take a very long time for the monolithic subsystem of the hybrid architecture to learn a policy that can perform better than the modular architecture. Furthermore, the modular system may actually be quite close to optimal. In particular, even though we used a relatively weak temporal discount ($\gamma = 0.95$) in most cases a greedy (or nearest neighbor) strategy is optimal. Only when there are very weak temporal discounts ($\gamma = 0.99$), high costs associated with actions, or goals that are mutually coupled (as in the example given in the previous section) does the nearest neighbor differ substantially from optimal. Thus, even if the simulations are extended significantly beyond 100,000 it is unlikely that the hybrid will (for this class of problems) significantly outperform the modular system. This bodes well for the viability of the modular approach.

7 SUMMARY

Reinforcement is almost certain to play a role in the future development of autonomous robotic systems. However, the prominence of that role depends upon the extent to which it can be scaled to complex robot learning task. In this chapter we have considered control tasks that require the robot to coordinate behavior in order to accomplish a series of independent time varying goals. We have shown how these tasks can be formulated as Markov Decision Processes by augmenting the state space description with a set of goal activation inputs. Formulating the problem in this way leads to an exponential growth in the state space size, and makes the straightforward application of Q-learning intractable. To mitigate this scaling problem, we have proposed a modular architecture, which decomposes the task by goals. Each goal is associated with a control module, whose objective is to learn to achieve that goal. Learning is facilitated by reducing the state space from a single monolithic one, whose size is exponential in the number of goals, to a set of fixed sized state spaces (one per goal). Overall control in the modular architecture is the responsibility of an arbiter, who uses information encoded in the individual goal-modules as a resource for decision making. In general, computationally feasible arbitration strategies yield sub-optimal (but good) performance. Thus, the modular architecture trades off optimal performance in the limit for faster learning and good initial performance. A hybrid architecture that integrates both modular and monolithic components is also considered. Although this architecture is expected to achieve the best performance overall, for some classes of problems it is likely that near-optimal performance can be achieved with a modular system using a simple merging strategy.

Acknowledgements

This material is based on work supported by the National Science Foundation under grants IRI-9003841 and IRI-8903582. This work was also supported by ONR research grant N00014-90-J-1811, Air Force - Rome Air Development Center research contract F30602-91-C-0010, and Air Force research grant AFOSR-91-0108. The Government has certain rights in this material. Josh Tenenberg also wishes to thank Indiana University South Bend for its partial support of this research.

4

MEMORY-BASED REINFORCEMENT LEARNING: CONVERGING WITH LESS DATA AND LESS REAL TIME

Andrew W. Moore
Christopher G. Atkeson

Artificial Intelligence Laboratory,
Massachusetts Institute of Technology, Cambridge, MA 02139

ABSTRACT

Prioritized Sweeping is a new algorithm for efficient prediction and control of stochastic Markov systems. Incremental learning methods such as Temporal Differencing and Q-learning have fast real time performance. Classical methods are slower, but more accurate, because they make full use of the observations. Prioritized Sweeping aims for the best of both worlds. It uses all previous experiences both to prioritize important dynamic programming sweeps and to guide the exploration of state-space. We compare Prioritized Sweeping with other reinforcement learning schemes for a number of different stochastic optimal control problems. It successfully solves large state-space real time problems with which other methods have difficulty.

1   INTRODUCTION

This paper provides a new algorithm to address the closely related problems of Markov prediction and reinforcement learning. Current, model-free, learning algorithms perform well relative to real time. Classical methods such as matrix inversion and dynamic programming perform well relative to the number of observations. The new algorithm, *Prioritized sweeping* seeks to achieve the best of both worlds. Its closest relation from conventional AI is the search scheduling technique of the $A^\star$ algorithm [Nilsson 1971]. Closely related research is being performed by

79

[Peng and Williams 1992] into a similar algorithm to prioritized sweeping, which they call Dyna-Q-queue.

We begin by providing a review of the problems and techniques in Markov prediction and control. More thorough reviews may be found elsewhere [Sutton 1988, Barto *et al.* 1989, Sutton 1990, Kaelbling 1990, Barto *et al.* 1991].

A discrete, finite Markov system has N_s *states*. Time passes as a series of discrete clock ticks, and on each tick the state may change. The probability of possible successor states is a function only of the current system state. The entire system can thus be specified by N_s and a table of transition probabilities.

$$\begin{matrix} q_{11} & q_{12} & \cdots & q_{1N_s} \\ q_{21} & q_{22} & \cdots & q_{2N_s} \\ \vdots & \vdots & & \vdots \\ q_{N_s 1} & q_{N_s 2} & \cdots & q_{N_s N_s} \end{matrix} \qquad (1)$$

where q_{ij} denotes the probability that, given we are in state i, we will be in state j on the next time step. Unsurprisingly, the table must satisfy $\sum_{j=1}^{N_s} q_{ij} = 1$ for every i.

The state-space of a Markov system is partitioned into two subsets: the non-terminal states N, and the terminal states T. Once a terminal state is entered, it is never left ($k \in T \Rightarrow q_{kk} = 1$).

A Markov system is defined as *absorbing* if from every non-terminal state it is possible to eventually enter a terminal state. We restrict our attention to absorbing Markov systems.

Let us first consider questions such as "starting in state i, what is the probability that I will eventually be absorbed by terminal state k?". Write this value as π_{ik}. All the absorption probabilities for terminal state k can be computed by solving the following set of linear equations. Assume that the non-terminal states are indexed by $1, 2, \ldots, N_n$ where N_n is the number of non-terminals.

$$\begin{aligned}
\pi_{1k} &= q_{1k} &+ q_{11}\pi_{1k} &+ q_{12}\pi_{2k} &+ \cdots &+ q_{1N_n}\pi_{N_nk} \\
\pi_{2k} &= q_{2k} &+ q_{21}\pi_{2k} &+ q_{22}\pi_{2k} &+ \cdots &+ q_{2N_n}\pi_{N_nk} \\
\vdots &\quad \vdots & & & &\quad \vdots \\
\pi_{N_nk} &= q_{N_nk} &+ q_{N_n1}\pi_{1k} &+ q_{N_n2}\pi_{2k} &+ \cdots &+ q_{N_nN_n}\pi_{N_nk}
\end{aligned} \tag{2}$$

When the transition probabilities $\{q_{ij}\}$ are known it is thus an easy matter to compute the eventual absorption probabilities. Machine learning can be applied to the case in which the transition probabilities are not known in advance, and all we may do instead is watch a series of state transitions. Such a series is normally arranged into a set of trials—each trial starts in some state and then continues until the system enters a terminal state.

Learning approaches to this problem have been widely studied. A recent contribution of great relevance is an elegant algorithm called *Temporal Differencing* [Sutton 1988].

1.1 The Temporal Differencing algorithm reviewed

We describe the discrete state-space case of the temporal differencing algorithm. TD can, however, also be applied to systems with continuous state spaces in which long term probabilities are represented by parametric function approximators such as neural networks [Tesauro 1991].

The prediction process runs in a series of epochs. Each epoch ends when a terminal state is entered. Assume we have passed through states $i_1, i_2, \ldots i_n, i_{n+1}$ so far in the current epoch. n is our age within the epoch and t is our global age. $i_n \to i_{n+1}$ is the most recently observed transition. Let $\hat{\pi}_{ik}[t]$ be the estimated value of π_{ik} after the system has been running for t state transition observations. The eventual absorption probabilities for the terminal states do not need to be estimated because they are known from the problem definition:

$$\forall k, k' \in T \text{ (the set of terminal states) and } \forall t$$
$$\hat{\pi}_{kk'}[t] = \begin{cases} 1 & \text{if } k = k' \\ 0 & \text{otherwise} \end{cases} \tag{3}$$

The simplest version of the TD algorithm is TD(0), which after observing the transition $i_n \rightarrow i_{n+1}$, updates the non-terminal absorption probabilities of i_n with the following rule.

for each $k \in T$ (the set of terminal states)
$$\hat{\pi}_{i_n k}[t+1] = (1-\alpha)\hat{\pi}_{i_n k}[t] + \alpha\hat{\pi}_{i_{n+1} k}[t] \qquad (4)$$

where α is a learning-rate constant, typically $\alpha = 0.05$. Thus, the eventual absorption estimates for the previous state are adjusted to move closer to the estimates for the new state.

The TD(λ) version of the algorithm generalizes this principle to allow updates not only to the previous state, but also earlier states in the recent epoch. The magnitude of the adjustment is reduced for older states, by an exponentially decaying factor λ^{age}. The full TD(λ) rule is:

for each $i \in N$ (the set of non-terminal states)
    for each $k \in T$ (the set of terminal states)
$$\hat{\pi}_{ik}[t+1] = \hat{\pi}_{ik}[t] + \alpha\left(\hat{\pi}_{i_{n+1}k}[t] - \hat{\pi}_{i_n k}[t]\right)\sum_{t'=0}^{t}\lambda^{t-t'}X_i(t') \qquad (5)$$

where
$$X_i(t') = \begin{cases} 1 & \text{if } i_{t'} = i \\ 0 & \text{otherwise} \end{cases} \qquad (6)$$

In practice there are computational tricks which require considerably less computation than the algorithm of Equation (5) but which compute the same values [Sutton 1988]. The first is to maintain the sum

$$e_i[t] = \sum_{t'=0}^{t}\lambda^{t-t'}X_i(t') \qquad (7)$$

for each state i. This is updated by $e_i[t+1] = X_i(t+1) + \lambda e_i[t]$. The TD algorithm then requires $O(N_t N_n)$ computation steps per real observation, where N_t is the number of terminal states and N_n is the number of non-terminals. An alternative optimization involves approximating Equation (5) by ignoring terms in the sum for which $\lambda^{t-t'}$ is below some threshold ϵ. Then only $N_a = \lceil \log(\lambda)/\log(\epsilon) \rceil$ increments need be made for each terminal state, and the computation steps per real observation are $O(N_t N_a)$.

If all non-terminal states can eventually reach a terminal state then
the expected value of the predictions is guaranteed to converge [Dayan
1992] The convergence of the expected values would not be very exciting
if nothing were said about the expected absoluté error. However, [Dayan
1992] also explains that if the α parameter of Equation (5) is allowed to
decay appropriately then

$$\lim_{t \to \infty} \hat{\pi}_{ik}[t] = \pi_{ik} \tag{8}$$

(remembering that $\hat{\pi}_{ik}[t]$ are the estimated absorption probabilities and
π_{ik} are the true values) with probability one.

1.2   The classical approach

The classical method proceeds by building a maximum likelihood model
of the state transitions. q_{ij} is estimated by

$$\hat{q}_{ij} = \frac{\text{Number of observations } i \to j}{\text{Number of occasions in state } i} \tag{9}$$

After $t + 1$ observations the new absorption probability estimates are
computed to satisfy, for each terminal state k, the $N_n \times N_n$ linear system

$$\hat{\pi}_{ik}[t + 1] = \hat{q}_{ik} + \sum_{j \in \text{ succs}(i) \cap N} \hat{q}_{ij} \hat{\pi}_{jk}[t + 1] \tag{10}$$

where succs(i) is the set of all states which have been observed as immediate successors of i and N is the set of non-terminal states. It is clear
that if the $\hat{q}_{ik}$ estimates were correct and if this sequence converged, it
would converge to the solution of Equation (2).

Notice that the values $\hat{\pi}_{ik}[t + 1]$ depend only on the values of $\hat{q}_{ik}$ after $t + 1$ observations—they are not defined in terms of the previous
absorption probability estimates $\hat{\pi}_{ik}[t]$. However, it is efficient to solve
Equation (10) iteratively. Let $\{\rho_{ik}\}$ be a set of intermediate iteration
variables containing intermediate estimates of $\hat{\pi}_{ik}[t + 1]$. What initial
estimates should be used to start the iteration? An excellent answer is
to use the previous absorption probability estimates $\hat{\pi}_{ik}[t]$.

The complete algorithm, performed once for every real-world observation, is shown in Figure 1. This is a *Gauss-Seidel* iterative algorithm, and

1. for each $i \in N$, for each $k \in T$,
   $$\rho_{ik} := \hat{\pi}_{ik}[t]$$

2. repeat
   2.1 $\Delta_{\max} := 0$

   2.2 for each $i \in N$
       for each $k \in T$
           Let $\rho_{\text{new}} = \hat{q}_{ik} + \sum_{j \in \text{ succs}(i)} \hat{q}_{ij} \rho_{jk}$ **in**
           $$\Delta := |\rho_{\text{new}} - \rho_{ik}|$$
           $$\rho_{ik} := \rho_{\text{new}}$$
           $$\Delta_{\max} := \max(\Delta_{\max}, \Delta)$$

   until $\Delta_{\max} < \epsilon$

3. for each $i \in N$, for each $k \in T$
   $$\hat{\pi}_{ik}[t+1] := \rho_{ik}$$

Figure 1.   Stochastic prediction with full Gauss-Seidel iteration.

is guaranteed to converge to a solution satisfying Equation (10) [Bertsekas and Tsitsiklis 1989]. If, according to the estimated transitions, all states can reach a terminal state, then this solution is unique. The inner loop ("for each $k \in T \cdots$") is referred to as a probability *backup* operation, and requires $O(N_t \mu_{\text{succs}})$ basic operations, where μ_{succs} is the mean number of observed stochastic successors.

Gauss-Seidel is an expensive algorithm, requiring $O(N_n)$ backups per real-world observation for the inner loop 2.2 alone. The absorption predictions before the most recent observation, $\hat{\pi}_{ik}[t]$, normally provide an excellent initial approximation, and only a very few iterations are required. However, when an "interesting" observation is encountered, for example a previously never-experienced transition to a terminal state, many iterations, perhaps more than N_n, are needed for convergence.

2 PRIORITIZED SWEEPING

Prioritized sweeping is designed to perform the same task as Gauss-Seidel iteration while using careful bookkeeping to concentrate all computational effort on the most "interesting" parts of the system. It operates in a similar computational regime as the Dyna architecture [Sutton 1990], in which a fixed, but non-trivial, amount of computation is allowed between each real-world observation. [Peng and Williams 1992] are exploring a closely related approach to prioritized sweeping, developed from Dyna and Q-learning [Watkins 1989].

Prioritized sweeping uses the Δ value from the probability update step 3.3 in the previous algorithm to determine which other updates are likely to be "interesting"—if the step produces a large change in the state's absorption probabilities then it is interesting because it is likely that the absorption probabilities of the predecessors of the state will change given an opportunity. If, on the other hand, the step produces a small change then there is little reason to bother with the predecessors. The predecessors of a state i are all those states i' which have, at least once in the history of the system, performed a one-step transition $i' \rightarrow i$.

If we have just changed the absorption probabilities of i by Δ, then the maximum possible one-processing-step change in predecessor i' caused by our change in i is $\hat{q}_{i'i}\Delta$. This value is the priority P of the predecessor i', and if i' is not currently on the priority queue it is placed there at priority P. If it *is* already on the queue, but at lower priority, then it is promoted.

After each real-world observation $i \rightarrow j$, the transition probability estimate $\hat{q}_{ij}$ is updated along with the probabilities of transition to all other previously observed successors of i. Then state i is promoted to the top of the priority queue so that its absorption probabilities are updated immediately. Next, we continue to process further states from the top of the queue. Each state that is processed may result in the addition or promotion of its predecessors within the queue. This loop continues for a preset number of processing steps or until the queue empties.

1. Promote state i_{recent} to top of priority queue.

2. While further processing allowed and queue not empty

   2.1 Remove the top state from the priority queue. Call it i

   2.2 $\Delta_{\max} = 0$

   2.3 for each $k \in T$
   $$\textbf{Let } \rho_{\text{new}} = \hat{q}_{ik} + \sum_{j \in \textbf{succs}(i) \cap N} \hat{q}_{ij} \hat{\pi}_{jk} \quad \textbf{in}$$
   $$\Delta := |\rho_{\text{new}} - \hat{\pi}_{ik}|$$
   $$\hat{\pi}_{ik} := \rho_{\text{new}}$$
   $$\Delta_{\max} := \max(\Delta_{\max}, \Delta)$$

   2.4 for each $i' \in \textbf{preds}(i)$
   $$P := \hat{q}_{i'i} \Delta_{\max}$$
   and if i' is not on queue, or P exceeds the current priority of i', then promote i' to new priority P.

Figure 2. The prioritized sweeping algorithm.

Thus if a real world observation is interesting, all its predecessors and their earlier ancestors quickly find themselves near the top of the priority queue. On the other hand, if the real world observation is unsurprising, then the processing immediately proceeds to other, more important areas of state-space which had been under consideration on the previous time step. These other areas may be different from those in which the system currently finds itself.

Let us look at the formal algorithm in Figure 2. On entry we assume the most recent state transition was from i_{recent}. We drop the $[t]$ suffix from the $\hat{\pi}_{ik}[t]$ notation.

The priority queue is implemented by a heap [Knuth 1973]. The cost of the algorithm is

$$O\left(\beta N_t(\mu_{\text{succs}} + \mu_{\text{preds}} \text{PQCOST}(N_n))\right) \tag{11}$$

basic operations, where at most β states are processed from the priority queue and $\texttt{PQCOST}(N)$ is the cost of accessing a priority queue of length N. For the heap implementation this is $\log_2 N$.

If all states can reach a terminal state, then the algorithm satisfies the conditions for asynchronous relaxation [Bertsekas and Tsitsiklis 1989] and convergence is guaranteed. Later, we will show empirically that convergence is relatively fast.

The memory requirements of learning the $N_s \times N_s$ transition probability matrix, where N_s is the number of states, may initially appear prohibitive, especially since we intend to operate with more than 10,000 states. However, we need only allocate memory for the experiences the system actually has, and for a wide class of physical systems there is not enough time in the lifetime of the system to run out of memory.

Similarly, the average number of successors and predecessors of states in the estimated transition matrix can be assumed $<< N_s$. A simple justification is that few real problems are fully connected, but a deeper reason is that for large N_s, even if the true transition probability matrix is not sparse, there will never be time to gain enough experience for the estimated transition matrix to not be sparse.

3   A MARKOV PREDICTION EXPERIMENT

Consider the 500 state Markov system depicted in Figure 3. The system has sixteen terminal states, depicted by white and black circles. The prediction problem is to estimate, for every non-terminal state, the long-term probability that it will terminate in a black, rather than a white, circle. The data available to the learner is a sequence of observed state transitions. Each learner was shown the same sequence.

Temporal differencing, the classical method, and prioritized sweeping were all applied to this problem. TD used parameters $\lambda = 0.25$ and $\alpha = 0.05$, which gave its best performance. The classical method was

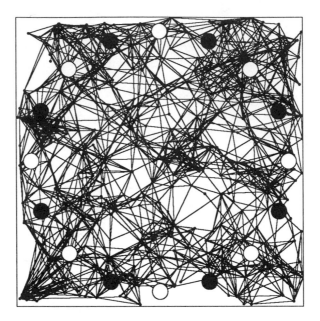

Figure 3. A 500-state Markov system. Each state has, on average, 5 stochastic successors.

required to compute up to date absorption probability estimates after every real-world observation. Prioritized sweeping was allowed five backups per real experience; it thus updated the $\hat{\pi}_{ik}$ estimates for the five highest priority states between each real-world observation. Each method was evaluated at a number of stages of learning, by stopping the real time clock and computing, the error between the estimated absorption probabilities $\hat{\pi}_{i,\ \text{WHITE}}$ and the true values $\pi_{i,\ \text{WHITE}}$. The RMS value over all states was recorded.

In Figure 4 we look at the RMS error plotted against the number of observations. After 100,000 experiences all methods are performing well; TD is the weakest but even it manages an RMS error of only 0.1. The classical method generally performs slightly better the prioritized sweeping.

In Figure 5 we look at a different measure of performance: plotted against real time cost. Here we see the great weakness of the classical technique. Performing the Gauss-Seidel algorithm of Figure 1 after each observation gives excellent predictions but is very time consuming,

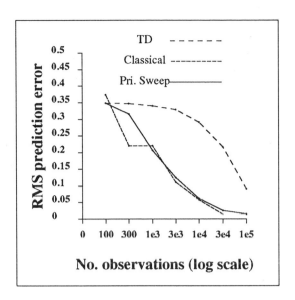

Figure 4. RMS prediction error between true absorption probabilities and predicted values $E = \sqrt{\frac{1}{N_n} \sum_{i=0}^{N_n} \left(\pi_{iW} - \hat{\pi}_{iW}\right)^2}$ graphed plotted against number of data-points observed.

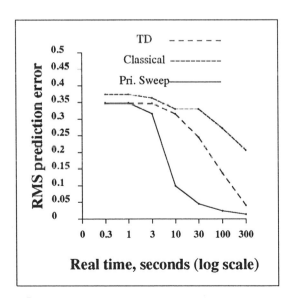

Figure 5. RMS prediction error between true absorption probabilities and predicted values graphed against real time, in seconds, running the problem on a Sun-4 workstation.

and after 300 seconds there has only been time to process a few hundred observations. After the same amount of time, TD has seen almost half a million observations, which it has been able to process quickly. Prioritized sweeping performs best relative to real time. It takes approximately ten times as long as TD to process each observation but because the data is used more effectively, convergence is superior.

This example has shown the general theme of this paper. Model-free methods perform well in real time but make weak use of their data. Classical methods make good use of their data but are often impractically slow. Techniques such as prioritized sweeping are interesting because they may be able to achieve both.

4 LEARNING CONTROL OF MARKOV DECISION TASKS

Let us consider a related stochastic prediction problem, which bridges the gap between Markov prediction and control. Suppose the system gets rewarded for entering certain states and punished for entering others. Let the reward of the ith state be r_i. An important quantity is then the *Expected discounted reward-to-go* of each state. This is an infinite sum of expected future rewards, with each term supplemented by an exponentially decreasing *discount factor*. The expected discounted reward-to-go is

$$
\begin{aligned}
J_i^\star \quad = \quad & \text{(This reward)}+ \\
\gamma \quad & \text{(Expected reward in 1 time step)}+ \\
\gamma^2 \quad & \text{(Expected reward in 2 time steps)}+ \\
& \vdots \\
\gamma^k \quad & \text{(Expected reward in } k \text{ time steps)}+ \\
& \vdots
\end{aligned}
\tag{12}
$$

where γ is the discount factor.

For each i, $J_i^\star$ can be computed recursively as a function of its immediate successors.

$$
\begin{aligned}
J_1^\star &= r_1 + \gamma(q_{11}J_1^\star + q_{12}J_2^\star + \dots + q_{1N_s}J_{N_s}^\star) \\
J_2^\star &= r_2 + \gamma(q_{21}J_2^\star + q_{22}J_2^\star + \dots + q_{2N_s}J_{N_s}^\star) \\
\vdots\quad &\quad\vdots \qquad\qquad\qquad\qquad\qquad\qquad\qquad\vdots \\
J_{N_s}^\star &= r_{N_s} + \gamma(q_{N_s1}J_1^\star + q_{N_s2}J_2^\star + \dots + q_{N_sN_s}J_{N_s}^\star)
\end{aligned}
\tag{13}
$$

which is another set of linear equations that may be solved if the transition probabilities q_{ij} are known. If they are not known, but instead a sequence of state transitions and r_i observations is given, then slight modifications of TD, the classical algorithm, and prioritized sweeping can all be used to estimate $J_i^\star$.

Markov decision tasks are an extension of the Markov model in which, instead of passively watching the state move around randomly, we are able to influence it.

Associated with each state, i, is a finite, discrete set of actions, actions(i). On each time step, the controller must choose an action. The probabilities of potential next states depend not only on the current state, but also on the chosen action. We use the notation q_{ij}^a for the probability that we move to state j, given that we have commenced in state i and applied action a.

A *policy* is a mapping from states to actions. If the controller chooses actions according to a fixed policy then it behaves like a Markov system. The expected discounted reward-to-go can then be defined and computed in the same manner as Equation (13).

If the goal is large reward-to-go, then some policies are better than others. An important result from the theory of Markov decision tasks tells us that there always exists at least one policy which is *optimal* in the following sense. For every state, the expected discounted reward-to-go using an optimal policy is no worse than that from any other policy.

Furthermore, there is a simple algorithm for computing both an optimal policy and the expected discounted reward-to-go of this policy. The algorithm is called *Dynamic Programming* [Bellman 1957]. It is based on the following relationship known as *Bellman's optimality equation* which holds between the optimal expected discounted reward-to-go at different states.

$$J_i^\star = \max_{a \in \ \textbf{actions}(i)} \left(r_i + \gamma(q_{i1}^a J_1^\star + q_{i2}^a J_2^\star + \cdots + q_{iN_s}^a J_{N_s}^\star) \right) \qquad (14)$$

A very important question for machine learning has been how to obtain an optimal, or near optimal, policy when the q_{ij}^a values are not known in advance. Instead, a series of actions, state transitions, and rewards is observed. A critical difference between this problem and the Markov prediction problem of the earlier sections is that the controller now affects which transitions are seen, because it supplies the actions.

The question of learning such systems is studied by the field of *reinforcement learning*, which is also known as "learning control of Markov decision tasks". Early contributions to this field were Samuel's checkers player [Samuel 1959] and the BOXES system of [Michie and Chambers 1968]. [Fu 1970] describes many other early approaches. Even systems which may at first appear trivially small, such as the two armed bandit problem [Berry and Fristedt 1985] have promoted rich and interesting work in the statistics community.

The technique of gradient descent optimization of neural networks in combination with approximations to the policy and reward-to-go (called, then, the "adaptive heuristic critic") was introduced by [Barto *et al.* 1983]. In [Kaelbling 1990] several applicable techniques were introduced, including the *Interval Estimation algorithm*. [Watkins 1989] introduced an important model-free asynchronous Dynamic Programming technique called Q-learning. Sutton has extended this further with the Dyna architecture [Sutton 1990]. An excellent review of the entire field may be found in [Barto *et al.* 1991]. The following references also provide reviews of the recent reinforcement learning literature: [Sutton 1988, Barto *et al.* 1989, Sutton 1990, Kaelbling 1990].

4.1   Prioritized sweeping for learning control of Markov decision tasks

The main differences between this case and the previous application of prioritized sweeping are

1. We need to estimate the optimal discounted reward-to-go, $J^\star$, of each state, rather than the eventual absorption probabilities.

2. Instead of using the absorption probability backup Equation (9), we use Bellman's equation [Bellman 1957, Bertsekas and Tsitsiklis 1989]:

$$J_i^\star = \max_{a \in \text{ actions}(i)} \left(\hat{r}_i^a + \gamma \times \sum_{j \in \text{ succs}(i,a)} \hat{q}_{ij}^a J_j^\star \right) \qquad (15)$$

where $J_i^\star$ is the optimal discounted reward starting from state i, γ is the discount factor, $\text{actions}(i)$ is the set of possible actions in state i, and $\hat{q}_{ij}^a$ is the maximum likelihood estimated probability of moving from state i to state j given that we have applied action a. The estimated immediate reward, $\hat{r}_i^a$, is computed as the mean reward experienced to date during all previous applications of action a in state i.

3. The rate of learning can be affected considerably by the controller's exploration strategy.

A detailed description of prioritized sweeping in conjunction with Bellman's equation, which would involve great repetition of the stochastic prediction case, is omitted. We will look instead at point 3 above.

The question of how best to gain useful experience in a Markov decision task is interesting because the formally correct method, when given a prior probability distribution over the space of Markov decision tasks, is unrealistically expensive. The formal solution uses dynamic programming over the space of all possible intermediate exploration scenarios and so is computationally exponential in all of (i) the number of time steps for which the system is to remain alive (ii) the number of states in the system, and (iii) the number of actions available [Berry and Fristedt 1985].

An exploration heuristic is thus required. [Kaelbling 1990] and [Barto *et al.* 1991] both give excellent overviews of the wide range of heuristics which have been proposed.

We use the philosophy of *optimism in the face of uncertainty*, a method successfully demonstrated by two earlier investigations:

- **Interval Estimation.** The IE algorithm of [Kaelbling 1990] uses statistics to estimate the expected reward, and spread of rewards, and chooses actions according to (for example) the 95th best percentile in the probability distribution of outcomes.

- **Exploration Bonus.** The exploration bonus technique in Dyna [Sutton 1990] increases the reward estimate for states which have not been visited for a long time.

A slightly different heuristic to either of these is used with the prioritized sweeping algorithm. This is because of minor problems of computational expense for IE and the instability of the exploration bonus in large state-spaces.

The slightly different optimistic heuristic is as follows. In the absence of contrary evidence, any action in any state is assumed to lead us directly to a fictional absorbing state of permanent large reward r^{opt}. The amount of evidence to the contrary which is needed to quench our optimism is a system parameter, T_{bored}. If the number of occurrences of a given state-action pair is less than T_{bored}, we assume that we will jump to fictional state with subsequent long term reward $r^{\text{opt}} + \gamma r^{\text{opt}} + \gamma^2 r^{\text{opt}} + \ldots = r^{\text{opt}}/(1-\gamma)$. If the number of occurrences is not less than T_{bored}, then we use the true, non-optimistic, assumption. Thus the optimistic reward-to-go estimate $J^{\text{opt}\star}$ is

$$
J_i^{\text{opt}\star} = \max_{a \in \text{ actions}(i)} \begin{cases} r^{\text{opt}}/(1-\gamma) & \text{if } n_i^a < T_{\text{bored}} \\ \hat{r}_i^a + \gamma \times \sum_{j \in \text{ succs}(i,a)} \hat{q}_{ij}^a J_j^{\text{opt}\star} & \text{otherwise} \end{cases} \tag{16}
$$

where n_i^a is the number of times action a has been tried to date in state i. The important feature, identified in [Sutton 1990], is the *planning to explore* behavior caused by the appearance of the optimism on both sides of the equation. Consider the situation in Figure 6. The top left hand corner of state-space only looks attractive if we use an optimistic heuristic. The areas near the frontiers of little experience will have high $J^{\text{opt}\star}$, and in turn the areas near those have nearly as high $J^{\text{opt}\star}$. Therefore, if prioritized sweeping (or any other asynchronous dynamic programming method) does its job, from START we will be encouraged to go north towards the unknown instead of east to the best reward

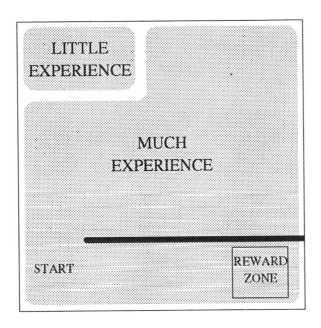

Figure 6. The state-space of a very simple path planning problem.

discovered to date. This behavior occurs because $J^{opt\star}$ appears on both sides of Equation (16).

The system parameter r^{opt} does not require fine tuning. It can be set to a gross overestimate of the largest possible reward, and the system will simply continue exploration until it has sampled all state-action combinations T_{bored} times.

The T_{bored} parameter, which defines how often we must try a given state-action combination before we cease our optimism, certainly does require forethought by the human programmer. If too small, we might overlook some low probability but highly rewarding stochastic successor. If too high, the system will waste time needlessly resampling already reliable statistics. Thus, the exploration procedure does not have full autonomy. This is, arguably, a necessary weakness of any non-random exploration heuristic. Dyna's exploration bonus contains a similar parameter in the relative size of the exploration bonus to the expected reward, and Interval Estimation has the parameter implicit in the optimistic confidence level.

Figure 7.   A three-DOF problem and the shortest solution path.

5   EXPERIMENTAL RESULTS

We begin with a task with a 3-d state-space quantized into 14,400 potential discrete states: guiding a planar rod through a maze by translation and rotation. There are four actions: move forwards 1 unit, move backwards 1 unit, rotate left 1 unit and rotate right 1 unit. In fact, the action takes us to the nearest quantized state after having applied the action. There are 20 × 20 position quantizations and 36 angle quantizations producing 14,400 states. The distance unit is 1/20th the width of the workspace and the angular unit is 10 degrees. The problem is deterministic but requires a long, very specific, sequence of moves to get to the goal. Figure 7 shows the problem, obstacles and shortest solution for our experiments.

Q-learning, Dyna-PI+, Optimistic Dyna and prioritized sweeping were all tested. Dyna and prioritized sweeping were both allowed one hundred Bellman's equation processing steps on each iteration. Two versions of Dyna were tested:

Technique	Experiences to converge	Real time to converge
Q	never	
Dyna-PI+	never	
Optimistic Dyna	55,000	1500 secs
Prioritized Sweeping	14,000	330 secs

Table 1. Performance on the deterministic rod-in-maze task. Both Dynas and prioritized sweeping were allowed 100 backups per experience.

1. Dyna-PI+ is the original Dyna-PI from [Sutton 1990], supplemented with the exploration bonus ($\epsilon = 0.001$) from the same paper.

2. Dyna-opt is the original Dyna-PI supplemented with the same T_{bored} optimistic heuristic that is used by prioritized sweeping.

Table 1 shows the number of observations before convergence. A trial was defined to have converged by a given time if no subsequent sequence of 1000 decisions contained more than 2% suboptimal decisions. The test for optimality was performed by comparison with the control law obtained from full dynamic programming using the true simulation.

Q and Dyna-PI+ did not even travel a quarter of the way to the goal, let alone discover an optimal path, within 200,000 experiences. It is possible that a very well-chosen exploration bonus would have helped Dyna-PI+ but in the four different experiments we tried, no value produced stable exploration.

Optimistic Dyna and prioritized sweeping both eventually converged, with the latter requiring a third the experiences and a fifth the real time. Not all potential state-action pairs were tried because many were unreachable from the start state. As a result, a stage was reached when no experienced state had any unexperienced actions and so there was no further optimism despite the fact that not all potential states had been visited.

When 2000 backups per experience were permitted, instead of 100, then both optimistic Dyna and prioritized sweeping required fewer experiences to converge. Optimistic Dyna took 21,000 experiences instead of 55,000 but took 2,900 seconds—almost twice the real time. Prioritized sweeping took 13,500 instead of 14,000 experiences—very little improvement, but it used no extra time. This indicates that for prioritized sweeping, 100 backups per observation is sufficient to make almost complete use of its observations. We conjecture that even full dynamic programming after each experience (which would take days of real time) would do little better.

We also consider a discrete maze world with subtasks, of the kind invented by [Singh 1991]. It consists of a maze and extra state information dependent on where you have visited so far in the maze. We use the example in Figure 8. There are 263 cells, but there are also four binary flags appended to the state, producing a total of $263 \times 16 = 4208$ states. The flags, named A, B, C and X, are set whenever the cell containing the corresponding letter is passed through. All flags are cleared when the start state (in the bottom left hand corner) is entered. A reward is given when the goal state (top right) is entered, only if flags A, B and C are set. Flag X provides further interest. If X is clear, the reward is 100 units. If X is set, the reward is only 50 units. There are four actions, corresponding to the four nearest neighbors of cell. The move only takes place if the destination cell is unblocked. Some experiments used stochastic maze dynamics in which cell transitions were randomly altered to a different neighbor 50% of the time.

Prioritized sweeping was tried with both the deterministic and stochastic maze dynamics. In both cases it found the globally optimal path through the three good flags to the goal, avoiding flag X. The deterministic case took 19,000 observations and twenty minutes of real time. The stochastic case required 120,000 observations and two hours of real time.

In these experiments, no information regarding the special structure of the problem was available to the learner. For example, knowledge of the cell at coordinates $(7, 1)$ with flag A set had no bearing on knowledge of the cell at coordinates $(7, 1)$ with A clear. If we told the learner that cell transitions are independent of flag settings then the convergence rate

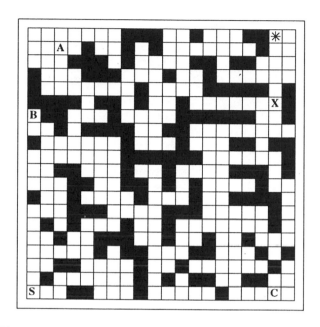

Figure 8. A maze with four binary bits of extra state.

would be increased considerably. A far more interesting possibility is the automatic discovery of such structure by inductive inference on the structure of the learned state transition matrix. See [Singh 1991] for current interesting work in that direction.

The third experiment is the familiar pole-balancing problem of [Michie and Chambers 1968, Barto *et al.* 1983]. The state-space of the cart is quantized at three equal levels for cart position, cart velocity, and pole angular speed. It is quantized at six equal levels for pole angle. The simulation used four real-valued state variables, yet the learner was only allowed to base its control decisions on the current quantized state. There are two actions: thrust left $10N$ and thrust right $10N$. The problem is interesting because it involves *hidden state*—the controller believes the system is Markov when in fact it is not. This is because there many possible values for the real-valued state variables in each discretized box, and successor boxes are partially determined by these real values, which are not given to the controller. The task is defined by a reward of 100 units for every state except one absorbing state corresponding to a crash, which receives zero reward. A discount factor of 0.999 is used. If the simulation contains no noise, or a very small amount (0.1% added to the simulated thrust), prioritized sweeping very quickly (usually in under 1000 observations and 15 crashes) develops a policy which provides stability for approximately 100,000 cycles. With a small amount of noise (1%), stable runs of approximately 20,000 time steps are discovered after, on average, 30 crashes.

It is interesting that we are able to get away with lying to the controller about the Markov nature of the system. However, this may be due to special features of the quantized cart-pole problem.

6 DISCUSSION

This paper has been concerned with discrete state systems in which no prior assumptions are made about the structure of the state-space. Despite the weakness of the assumptions, we can successfully learn large stochastic tasks. However, very many problems do have extra known structure in the state-space, and it is important to consider how this knowledge can be used. By far the most common knowledge is smoothness — given two states which are in some way similar, in general their transition probabilities will be similar.

TD can also be applied to highly smooth problems using a parametric function approximator such as a neural network. This technique has recently been used successfully on a large complex problem, Backgammon, by [Tesauro 1991]. The discrete version of prioritized sweeping given in this paper could not be applied directly to Backgammon because the game has 10^{23} states, which is unmanageably large by a factor of at least 10^{10}. However, a method which quantized the space of board positions, or used a more sophisticated smoothing mechanism, might conceivably be able to compute a near-optimal strategy.

We are currently developing memory-based algorithms which take advantage of local smoothness assumptions. In these investigations, state transition models are learned by memory-based function approximators [Moore and Atkeson 1992]. Prioritized sweeping takes place over non-uniform tessellations of state-space, partitioned by variable resolution

kd-trees [Moore 1991]. We are also investigating the role of locally linear control rules and reward functions in such partitionings, in which instead of using Bellman's Equation (15) directly, we use local linear quadratic regulators (LQR) (see, for example, [Sage and White 1977]). It is worth remembering that, if the system is sufficiently linear, LQR is an extremely powerful technique. In a pole balancer experiment in which we used local weighted regression to identify a local linear model, LQR was able to create a stable controller based on only 31 state transitions!

Other current investigations which attempt to perform generalization in conjunction with reinforcement learning are [Mahadevan and Connell 1990] which investigates clustering parts of the policy, [Chapman and Kaelbling 1991] which investigates automatic detection of locally relevant state variables, and [Singh 1991] which considers how to automatically discover the structure in tasks such as the multiple-flags example of Figure 8.

6.1   Related work

Peng and Williams have concurrently been developing a closely related algorithm which they call Dyna-Q-queue [Peng and Williams 1992]. Where prioritized sweeping provides efficient data processing for methods which learn the state transition model, Dyna-Q-queue performs the same role for Q-learning [Watkins 1989], an algorithm which avoids building an explicit state-transition model. Dyna-Q-queue is also more careful about what it allows onto the priority queue: it only allows predecessors which have a predicted change ("interestingness" value) greater than a significant threshold δ, whereas prioritized sweeping allows everything above a minuscule change ($\frac{1}{10000}$ of the maximum reward) onto the queue. The initial experiments in [Peng and Williams 1992] consist of sparse, deterministic maze worlds of several hundred cells. Performance, measured by total number of Bellman's equation processing steps before convergence, is greatly improved over conventional Dyna-Q [Sutton 1990].

In other related work [Sutton 1990], Sutton identifies reinforcement learning with asynchronous dynamic programming and introduces the same computational regime as that used for prioritized sweeping. The notion of using an optimistic heuristic to guide search goes back to the $A^\star$ tree search algorithm [Nilsson 1971], which also motivated another aspect of prioritized sweeping: it too schedules nodes to be expanded according to an (albeit different) priority measure. More recently [Korf 1990] gives a combination of $A^\star$ and Dynamic Programming in the LRTA* algorithm. LRTA* is, however, very different from prioritized sweeping: it concentrates all search effort in a finite-horizon set of states beyond the current actual system state. Finally, [Lin 1991b] has investigated a simple technique which replays, backwards, the memorized sequence of experiences which the controller has recently had. Under some circumstances this may produce some of the beneficial effects of prioritized sweeping.

7   CONCLUSION

Our investigation shows that prioritized sweeping can solve large state-space real time problems with which other methods have difficulty. Other benefits of the memory-based approach, described in [Moore and Atkeson 1992], allow us to control forgetting in potentially changeable environments and to automatically scale state variables. Prioritized sweeping is heavily based on learning a world model and we conclude with a few words on this topic.

If a model of the world is not known to the human programmer in advance then an adaptive system is required, and there are two alternatives:

Learn a model and from this develop a control rule.	Learn a control rule without building a model.

Dyna and prioritized sweeping fall into the first category. Temporal differences and Q-learning fall into the second. Two motivations for *not* learning a model are (i) the interesting fact that the methods do, nevertheless, learn, and (ii) the possibility that this more accurately simulates some kinds of biological learning [Sutton and Barto 1990]. However, a third advantage which is sometimes touted—that there are computational benefits in not learning a model—is, in our view, highly dubious. A common argument is that with the real world available to be sensed directly, why should we bother with less reliable, learned internal representations? The counterargument is that even systems acting in real time can, for every one real experience, sample millions of mental experiences from which to make decisions and improve control rules.

Consider a more colorful example. Suppose the anti-model argument was applied by a new arrival at a university campus: "I don't need a map of the university—the university is its own map." If the new arrival truly mistrusts the university cartographers then there might be an argument for one full exploration of the campus in order to create their own map. However, once this map has been produced, the amount of time saved overall by pausing to consult the map before travelling to each new location—rather than exhaustive or random search in the real world—is undeniably enormous.

It is certainly justified to complain about the indiscriminate use of combinatorial search or matrix inversion prior to each supposedly real time decision. However, models need not be used in such an extravagant fashion. The prioritized sweeping algorithm is just one example of a class of algorithms which can easily operate in real time and derive great power from a model. In the 1990s, with Megabytes and Megaflops already cheap, and becoming even cheaper, the question should not be 'How can we minimize the amount of computation per control cycle?' Instead it should be 'How can we make best use of the available computing power?'

Acknowledgements

Thanks to Mary Soon Lee, Satinder Singh and Rich Sutton for useful comments on an early draft. Andrew W. Moore is supported by a Postdoctoral Fellowship from SERC/NATO. Support was also provided under Air Force Office of Scientific Research grant AFOSR-89-0500, an Alfred P. Sloan Fellowship, the W. M. Keck Foundation Associate Professorship in Biomedical Engineering, Siemens Corporation, and a National Science Foundation Presidential Young Investigator Award to Christopher G. Atkeson.

RAPID TASK LEARNING
FOR REAL ROBOTS

Jonathan H. Connell
Sridhar Mahadevan

IBM T. J. Watson Research Center,
Box 704, Yorktown Heights, NY 10598

ABSTRACT

For learning to be useful on real robots, whatever algorithm is used must converge in some "reasonable" amount of time. If each trial step takes on the order of seconds, a million steps would take several months of continuous run time. In many cases such extended runs are neither desirable nor practical. In this chapter we discuss how learning can be speeded up by exploiting properties of the task, sensor configuration, environment, and existing control structure.

1 INTRODUCTION

Automatic programming of robots using machine learning is an attractive area for study. Of the several possible approaches, reinforcement learning is particularly appealing since it can potentially translate an abstract specification of some task into a procedural program for actually accomplishing this objective. Unfortunately, most work in this area has been done on toy problems where the answers are simple, or on simulators where the agent can easily perform vast numbers of experiments. Yet it is very hard to guarantee a tight correspondence between a simulated environment and the real world. In the limit, consider the difficulties in using computer graphics to synthesize video inputs. In sum, to get robots to work in the real world, they must be trained (at least partially) in the real world.

Each step on a real robot requires a certain non-infinitesimal amount of time to perform. This delay is due to factors such as the inertia of the robot, the maximum torque its motors can produce, and the speed of sound. In some cases it may be acceptable to allow a single robot to spend a long time learning its task and then copy the learned control rules to a collection of identical robots. However, this solution is not viable if the robots are to be placed in substantially different environments, or each robot has to learn new tasks every day. Thus, in many cases there is a real need for learning algorithms implementable on real robots that can converge in a few thousand steps or less.

In this chapter we will explore several ways in which the convergence rate of a learning algorithm can be improved by adding certain types of information. In particular, in Section 2 we explain why complex reward functions speed up learning, and how a behavior-based architecture [Brooks 1986, Connell 1990] can be used to avoid local maxima in these functions. In Section 3 we increase the size of the input representation to allow for better discrimination at long ranges. To counteract the complexity of producing generalization over this new representation we show how to take advantage of the local spatial structure of the sensors. In Section 4 we use the same input representation but amortize the cost of learning across several different tasks. We do this by building a model of the world which tells how the robot's actions change its sensory input irrespective of the current objective. Finally, in Section 5 we describe a mapping system which exploits its own low-level behaviors to handle most of the variability in its world. This approach reduces the learning task to the acquisition of only a small amount of connectivity information about the environment. All these systems can learn their assigned tasks in real-world run times of substantially less than a day.

2   BEHAVIOR-BASED REINFORCEMENT LEARNING

Teaching a robot to carry out a task essentially reduces to learning a mapping from perceived states to desired actions. The preferred mapping is one that maximizes the robot's performance at the particular task. Reinforcement learning [Sutton 1991] is an area of machine learning that studies the problem of inferring a control policy that optimizes a fixed cost function. Traditionally, reinforcement learning has focussed on simulated tasks such as pole balancing [Michie and Chambers 1968, Barto *et al.* 1983], although more recently it has been successfully applied to more interesting domains such as backgammon [Tesauro 1991]. However, in these applications the learner receives a very delayed reward only after completing the task (for example, after dropping the pole or winning a game). Whitehead [Whitehead 1991a] has theoretically shown that learning converges very slowly with such infrequent rewards. This theoretical result, which agrees with our intuitions, has also been confirmed in many experimental studies.

Thus, in this section our principal concern is improving the convergence of reinforcement learning by increasing the feedback given to the learner. A naive approach would be to write a more complex reward function that returns non-zero values at many intermediate states prior to completing a task. However, such an approach suffers from the problem of local maxima – the learner settles into a very suboptimal policy that maximizes just one of the subterms of the reward function. Instead, we show how the local maxima problem can be avoided if a modular control architecture is used. Roughly speaking, each subterm of the reward function can be assigned to a distinct module. Each module maintains a separate control function, and conflicts among modules are resolved using a hardwired arbitration strategy. A detailed description of this approach is given in [Mahadevan and Connell 1992].

2.1   The Box-Pushing Task

To explore ways in which robots can learn, we have to pick a sample task that is rich enough that it may potentially lead to practical use, yet simple enough to formulate and study. One such task is having a robot push boxes across a room. One can view this as a simplified version of a task carried out by a warehouse robot moving packing cartons around from one location to another.

The robot used in our experiments (shown in Figure 1) is built on a small, 12" diameter, 3-wheeled base from RWI. The sensors are mounted on

Figure 1.   The robot used in these experiments.

top and always face in the direction of travel. For our experiments, we limit the motion of the vehicle to one of 5 different actions. The robot either moves forward, or turns left or right in place by two different angles (22 degrees or 45 degrees).

OBELIX's primary sensory system is an array of 8 sonar units (see Figure 2). Each sensor in the array has a field of view of roughly 20 degrees. The individual sonar units are arranged in an orthogonal pattern. There are 4 sonars looking toward the front and 2 looking toward each side. For the purposes of the experiments described here, we use only two range bins for each sonar. One extends from 9" to 18" (NEAR) and another covers the distance between 18" and 30" (FAR).

There are also two secondary sources of sensory information that have higher semantic content than the general-purpose sonar array. There is an infra-red (IR) detector which faces straight forward and is tuned to a response distance of 4". This sensor provides a special bit called BUMP since it only comes on when something is right against the front

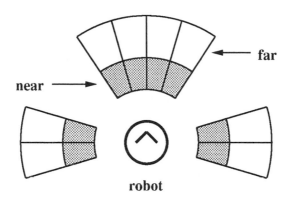

Overhead View

Figure 2. Sixteen sonar echo range bins provide the input vector.

of the robot. The robot also monitors the motor current being used for forward motion. If this quantity exceeds some fixed threshold, another special bit, STUCK, is turned on.

To summarize, 18 bits of information are extracted from the sensors on the robot. Of these, 16 bits of information come from the 8 sonar sensors (1 bit from each NEAR range and 1 bit from each FAR range). There is also 1 bit of BUMP information, and 1 bit of STUCK information. The 18 bits generate a total sensor space of about a quarter million states. It is the job of the learning algorithm to decide which of the 5 actions to take in each of these states.

2.2 A Monolithic Controller

Let us first design an agent who learns the box pushing task in its entirety. That is, we create a *monolithic* learner by defining a single module – box pusher – that is active all the time. The reward function for the monolithic learner is given in Figure 3. The single module is given a reward of 1 when it pushes a box – that is, it was bumped in two successive states while going forward, and was not stuck in the second state –

```
Monolithic_reward (old_state,action,new_state):
begin
      IF action = forward                    % went forward
            and BUMP(old_state)              % bumped before
            and BUMP(new_state)              % bumped now
            and ¬STUCK(new_state)            % not stuck now
      THEN return 1                          % reward robot
      ELSE return 0                          % default is no reward
end
```

Figure 3. Reward Function for Monolithic Learner

and is given a reward of 0 otherwise. Note that the monolithic reward function can also be used to measure the percentage of time the robot was actually pushing a box.

We use a variation of Q learning [Watkins 1989] to solve the temporal credit assignment problem of deciding which actions in a given sequence were chiefly responsible for a received reward. This technique uses a utility function $Q(x, a)$ across states (x) and actions (a) to represent a control mapping. The utility of doing an action a in a state x, or $Q(x, a)$, is defined as the sum of the immediate payoff or reward r plus the utility $E(y)$ of the state y resulting from the action *discounted* by a parameter γ between 0 and 1. That is, $Q(x, a) = r + \gamma E(y)$. The utility of a state x, in turn, is defined as the maximum of the utilities of doing different actions in that state. That is, $E(x) = \text{maximum } Q(x, a)$ over all actions a. During learning, the stored utility values $Q(x, a)$ have not yet converged to their final value (that is, to $r + \gamma E(y)$). Thus, the difference between the stored values and their final values gives the error in the current stored value. In particular, Q learning uses the following rule to update stored utility values.

$$Q(x, a) \leftarrow Q(x, a) + \beta(r + \gamma E(y) - Q(x, a))$$

Thus, the new Q value is the sum of the old one and the error term multiplied by a parameter β, between 0 and 1. The parameter β controls the rate at which the error in the current utility value is corrected. In our experiments the learning rate β, and the discount factor γ were set at 0.5 and 0.9, respectively.

In general, the best action a to perform in a given state x is the one that has the highest utility $Q(x, a)$. So essentially the Q values implement a state-action control table for a given task. However, by forcing the robot to always select the highest Q-value action prevents it from ever exploring new actions. Therefore, some percentage of the time, a random action is chosen to ensure that all states in the state space will eventually be explored. Exploring all states is a necessary condition for Q learning to converge [Watkins 1989]. We have experimentally found 10% to be a good compromise between exploratory and goal-directed activity. Usually, once the robot has learned a good control policy, it can easily compensate for occasional deviations from its strategy caused by random actions.

Q learning solves the temporal credit assignment problem, but not the structural credit assignment problem of inferring rewards in novel states. We use statistical clustering to propagate reward values across states. The robot learns a set of clusters for each action that specify the class of states in which the action should or should not be done. More formally, a cluster is a vector of probabilities $< p_1, ..., p_n >$, where each p_i is the probability of the ith state bit being a 1. Each cluster has associated with it a Q value indicating its worth, and a count representing the number of states that matched it. For example, the robot might aggregates all states where going forward gave rise to a positive Q value into a cluster. Clusters are extracted from instances of states, actions, and rewards that are generated by the robot exploring its task environment.

A state s is considered an instance of a cluster if two conditions are satisfied. One, the product of the probabilities p_i or $(1 - p_i)$ – depending on whether the ith bit of state s is a 1 or a 0 – is greater than some threshold ϵ. Two, the absolute difference between the Q values of the state and the cluster should be less than some threshold δ. If a state s matches a cluster, it is merged into the cluster by updating the cluster probabilities, and incrementing the cluster instance count.

New clusters are formed in two situations. If a state does not match any of the existing clusters, a new cluster is created with the current state being its only instance. Alternatively, two clusters can be merged into one "supercluster" if the Euclidean "distance" between the clusters (treating the clusters as points in n dimensional probability space) is less than some threshold ρ, and the difference between their Q values is less than δ.

2.3   A Behavior-based Controller

We spent a few frustrating months trying to get our mobile robot to push boxes using the simple monolithic reinforcement learning structure outlined above. This approach failed for a number of reasons. If a simple reward function was used, the robot got rewards very infrequently, and learned very slowly. Increasing the frequency of rewards by using a more complex reward did not speed up the learning. Instead, the robot got stuck in local minima – for example, OBELIX would either avoid coming in contact with any object, or it would try to push everything in sight! The perceptual aliasing problem [Whitehead and Ballard 1990] also haunted us – boxes often look like walls in sonar images. Finally, we could not find effective ways to encode state history information. Encoding histories by adding previous states to the current sensor readings tended to slow the learning even further. As a consequence, the robot never exhibited state history dependent behaviors – it would often turn back to push a wall forgetting that a few steps back it had tried to push it and failed.

Eventually we realized that the box pushing task actually involves several distinct subtasks, and that having the robot learn each subtask separately would solve many of these problems. The same approach is taken in behavior-based robotics (for example, [Brooks 1986, Brooks 1989, Connell 1989, Connell 1990]). Conceptually, the box pushing task involves three subtasks. First, the robot needs to find potential boxes and discriminate them from walls and other obstacles, which is quite difficult to do using sonar or infra-red detectors. Second, the robot needs to be able to push a box across a room, which is tricky because boxes tend to slide and rotate unpredictably. Finally, the robot needs to be able to recover from stalled situations where it has either pushed a box to a corner, or has attempted to push an immovable object like

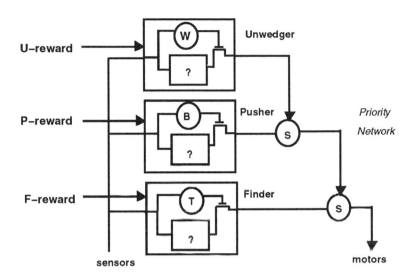

Figure 4. A behavior-based control architecture for the box-pushing task.

a wall. Our approach will be to learn each of these subtasks as a distinct *reactive* behavior. By reactive, we mean that we base the control decision on only the currently perceived sensory information. Figure 4 illustrates the overall structure of a behavior-based robot that follows this decomposition of the task.

Our intention is to have OBELIX learn the above three tasks in parallel, using three copies of the learning algorithm. The ordering on the three behaviors enables OBELIX to determine the dominant behavior in any situation if there is a conflict (**S** indicates suppression). In addition to the priority structure, each module has its own "applicability predicate" (**W, B, T**) that gates the output of the related transfer function (**?**). These functions combined with the priorities are essentially a "sketch" of a plan for pushing a box. The learning system's job is to flesh out the details by coming up with suitable transfer functions for the three modules. We can visualize the control flow among modules in a box pushing robot as shown in Figure 5. The robot always starts in the finder module (labeled **F** in the figure). If it is bumped, the pusher module (labeled **P**) turns on. The pusher module continues to control the robot if the robot continues to be bumped. If it becomes stuck, control transfers to the unwedger module (labeled **U**). The timeout refers

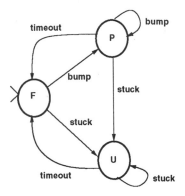

Figure 5. Equivalent control flow between modules.

to the number of steps that a module continues to be applicable after its initial triggering condition is no longer true.

2.4   Performance

To collect performance data using the real robot OBELIX, a special "playpen" was created consisting of a portion of the robotics laboratory at IBM Hawthorne. Large packing cartons formed three walls of the playpen and a sheet rock wall formed the fourth. Two rectangular boxes were placed in the playpen for OBELIX to push around.

We used the following procedure to collect the experimental data. The learning runs were organized into 20 trials, each trial lasting for a 100 steps. Thus, the robot takes totally 2000 steps in each learning run. On our real robot, a learning run takes about 2 hours. At the beginning of each trial, the world is restored to its initial state. This includes resetting the boxes and the robot to their initial locations. The robot starts with a randomly chosen orientation at the beginning of each trial.

Figure 6 shows the results of using monolithic versus behavior-based controllers for this learning task. The performance of both the statistical clustering technique discussed previously and another method, called Weighted Hamming, are plotted here. In each case we show the cumulative average reward value for the learner. That is, the sum of all

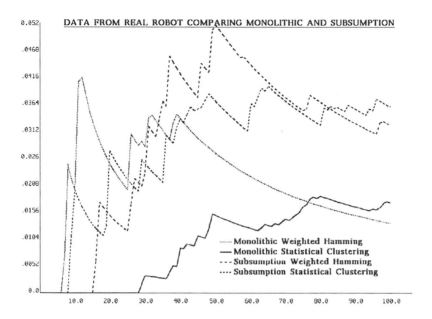

Figure 6. Learning speed in a subsumption versus monolithic architecture using a real robot. The horizontal axis is the percentage of the trial completed while the vertical axis is the fraction of steps so far during which the robot was pushing a box.

rewards received since the beginning divided by the number of actions taken. Even though the subsumption controller has three separate reward functions, we can still compute what the monolithic reward would have been for the overall system. This synthesized reward is graphed here. For reference, an agent that simply executes a random action at each time step yields a more or less constant reward of 0.01. By contrast, a static hand-coded program for the box-pushing task gives a flat reward of 0.04.

As can be seen, the subsumption system outperforms the monolithic system by about a factor of two. However, each system does learn something since the values shown are much better than those for the random agent. In fact, the subsumption approach is just about as good as the program we hand-coded for the task. Although the curves do not quite cross, the final average *performance* value for the subsumption learner is actually higher than shown. What has been graphed is the average *cumulative* reward. This running average has been depressed by the inclusion of the initial period of poor performance.

We can also look at how well the two approaches learn the individual subtasks involved in box pushing. Here we use the three reward functions provided to the subsumption learner. As before, although the monolithic system is not provided with these signals, given the sensor pattern and action for each step we can synthesize what they would have been. In Table 1 we report the percentage improvement for each subtask. This is calculated by first taking the final cumulative reward average (that is, the last point in a graph like Figure 6) and subtracting the worst possible reward that could have been received. This value is then normalized by dividing it by the difference between the best and worst possible rewards that could have been received. The actual numbers reported here are from a different structural credit assignment technique called Weighted Hamming [Mahadevan and Connell 1992]. For comparison, the Subsumption Clustering technique achieved improvements of 29%, 60%, and 72% on the subtasks of Finding, Pushing, and Unwedging respectively.

Technique	Finder	Pusher	Unwedger
Handcoded Agent	31%	57%	73%
Subsumption WH	15%	55%	74%
Monolithic WH	7%	46%	67%
Random Agent	8%	30%	68%

Table 1. Average performance at end of learning run for real data.

The table indicates that the subsumption approach is reasonably successful at learning to push boxes and unwedge from stalled states – the performance comes close to or exceeds that of the handcoded agent. The subsumption learner is clearly less successful at learning to find boxes. This is what we expected, given the difficulty of distinguishing boxes from obstacles using sonar. Also, while both the subsumption and monolithic controller receive good scores for recovering from stalled states, the random agent does nearly as well. This result suggests that simply thrashing can often extricate the robot from these situations.

Comparing the rows in table reveals that the monolithic learner really only learned to push a box. In terms of finding boxes and unwedging from stalled states, the monolithic learner is only as good as the random agent. Thus, the reason that the monolithic learner performs the overall task only half as well as the subsumption learner is mainly because it never learned to find boxes. Note that the monolithic learner receives rewards only when it pushes a box, a fairly rare event during the learning run. In fact, with such a reward the early stages of the learning are essentially a random walk [Whitehead 1991a]. The monolithic system's poor performance at finding boxes suggests that, over the limited length of a learning run, the learning technique used was unable to resolve the long-term temporal credit assignment problem involved in approaching potential boxes.

3 EXPLOITING LOCAL SPATIAL STRUCTURE

In the previous section our robot learned the box-pushing task using an input representation consisting of 18 binary values. However, there is not enough information in this representation to allow the robot to reliably distinguish boxes from walls until the robot is adjacent to the object in question. When the robot is fairly far away, neither type of environmental feature can be registered properly. To overcome this sensory limitation, in this section we extend the robot's sensory input to be an array representing an overhead view of the robot and its surroundings. This array consists of a number of uniformly sizes cells, each of which contains a number telling how confident the robot is that this cell is occupied by some material object. We typically use 12 by 12 arrays where the squares are 6 inches on a side.

Although richer sensors enable better discrimination, they also make learning more difficult. In particular, learning a control function in a reasonable amount of time from inputs of the size of 144 real numbers is a daunting prospect. The solution presented here was developed by Nitish Swarup, one of our summer students from MIT [Swarup 1991]. The idea is to essentially decouple the cells so that each cell gets to "vote" for a particular action with a strength proportional to its occupancy and a learned weight. We are assuming that superposition will work so that there are only 144 independent small problems to solve instead of one big problem. There is evidence that some animal behaviors are implemented in this manner [Arbib and House 1987, Manteuffel 1991].

We call full collection of cell weights a "Q-map". As the robot roams around, incoming state inputs are multiplied cell-wise with the Q-map for each action to produce a net utility value for doing that action. The action with the highest utility becomes the action selected by the robot. The reinforcement learning algorithm is set up to strengthen the weights in an action's Q-map if the corresponding cell in the input state was occupied when a positive reward was received. The weights of any cells which were occupied when negative rewards were received are similarly decreased in value. For the box pushing example, the three modules would have separate Q-maps for each of the five actions, thus there would be 15 Q-maps altogether.

An example set of Q-maps is shown in Figure 7. The circle offset from the center represents the position of the robot while the interior arrowhead points in the direction of the robot's orientation. These Q-maps are associated with the pusher module in the behavior-based architecture shown earlier in Figure 4. The darkness of a cell is proportional to that cell's importance. Thus, solid black represents the most important cells, whereas solid white denotes the least significant cells in the Q-map. These grids makes intuitive sense: those sensory states with objects in front of the robot, for example, will produce high scores for the action "forward" due to the heavy positive weights in the spatially corresponding cells in the Q-map. The other action Q-maps do not produce such a high value because their "important" areas are concentrated elsewhere.

3.1   Scrolling Occupancy Grids

As mentioned before, the input to the robot consists of an array of scalar values representing how strongly the robot believes each patch of floor around it is occupied. These values range from 0 (empty) to 1 (full). This homogeneous spatial representation was originally developed by Moravec and Elfes [Moravec 1988, Moravec and Elfes 1985]. Unlike Elfes we do not maintain a full map of the world, only some small portion of it directly around the robot. We also eschew the formal probability

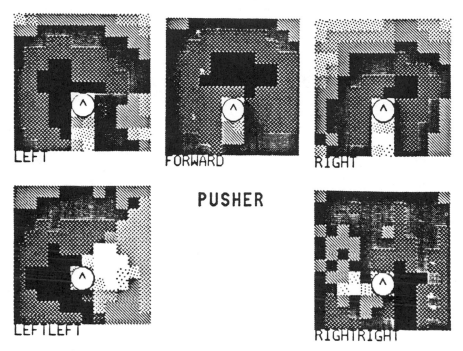

Figure 7. Q-maps learned for the actions in one module.

model and use an approximation instead (as described below). This simplification seems to be adequate for our purposes.

We build an occupancy grid by first plotting the sonar returns in an array of the same size as the map. Here we use the actual time to first echo, not range bins as in the previous section. Each sonar return generates an arc whose size and position are determined by the echo time combined with the sensor's location on the robot and its pointing direction. The width of each arc is always 20 degrees (the beam spread of the sonar). The cells near this arc are marked as likely to be occupied while the cells between this arc and the transducer position are marked as likely empty. The arcs farther from the robot are longer due the uncertainty in the exact angular position of the reflecting surface. For this reason, we give cells certainties starting with 0.8 for distances up to 18 inches, and a value derated inversely with distance beyond this. Empty cells are always given a certainty of -0.2.

These certainty values are used to alter the occupancy values in the grid from the previous time step. The old map is first shifted and rotated to compensate for the robot's own motion (as based on odometry). If

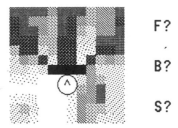

F?

B?

S?

Figure 8. A state description based on a scrolling occupancy grid as computed from simulated sonar readings. Three extra bits, F, B, and S, tell whether there is something in front of the robot, whether it is bumped, and whether it is stuck, respectively.

temporal fusion is not desired or this is the first map, all cells are initialized to 0.5 instead. Now, if the new certainty value for a cell is negative, the old occupancy value is diminished by the given percentage. If the certainty value is positive, the cell's occupancy is increased by adding in the specified percentage of the difference between the current value and 1. In this way new and old sonar scans are combined to yield a map centered on the robot's current position. The integration technique allows the robot to fill in blind spots in its sensory array by rotating. However, once objects fall off the edge of the map they are forgotten, even if the robot reverses its direction and re-enters the area it just left.

For the Q-map technique, we used a square occupancy grid consisting of 144 cells each of which was 6 inches on a side. For the real robot, special care was taken to accurately model the sensor geometry and viewing angles to give good fidelity at close ranges. We can also generate maps from an on-board scanning range-finder. However, for most of the work reported here we used sonar returns generated by the simulator which was used in some of our earlier work [Mahadevan and Connell 1992]. In general, the occupancy maps generated from these returns look better than those from the real robot. Figure 8 shows a map from the simulator. The robot has been moving for a while and integrating the sonar responses. It can be seen that the robot is directly against an obstruction and the grayish line on the right implies that it is likely in a corner.

3.2 Direct Sensory-Action Maps

As noted before, each module has a Q-map for each of the possible actions. However, not all the Q-maps are utilized at each step. There is additional control structure (the applicability predicates and priority arbitrator shown in Figure 4) that specifies which module is active at each time step. Only the Q-maps of the active module need be evaluated. Similarly, when a reward is received only the selected Q-map is updated. It would not make sense to reward or punish all the other maps which are not currently in control of the robot.

After the relevant module has been determined, the best action must be chosen. However, before multiplying the current scrolling occupancy grid by the stored Q-maps, the robot preprocesses this input data. A Gaussian filter is passed over the input in order to decrease the importance of radially-far state input cells. Intuitively, sensory systems should place greater importance on objects that are in the direct vicinity of the observer. Thus, each raw input cell of an incoming state representation is transformed to give

$$s_{i,j}^p = s_{i,j} e^{-\frac{d^2}{\tau}}$$

where $s_{i,j}$ represents the value of each raw state input cell before visual preprocessing. The parameter d is the distance, $\sqrt{(i - r_i)^2 + (j - r_j)^2}$, from the center of each state cell to the robot's position in the occupancy grid, (r_i, r_j). The depth acuity parameter τ controls the importance of physically distant cells. The preprocessing step encodes additional information about the domain by indirectly marking the radial distance of each cells. They are no longer just a collection of 144 unrelated numbers.

After radial preprocessing, the quantity below is calculated for each of the active Q-maps.

$$Q(s, a, m) = \frac{\sum_{j=0}^{11} \sum_{i=0}^{11} s_{i,j}^p G_{i,j,a,m}}{\sum_{j=0}^{11} \sum_{i=0}^{11} s_{i,j}^p}$$

The numerator of this expression is the sum of the probabilities in the current state occupancy map weighted by the corresponding conductance values, $G_{i,j,a,m}$, in the Q-map. The denominator normalizes the expression according to the sum of cell occupancy probabilities. The result of this expression is the current *Q-value*, $Q(s, a, m)$, computed by module m for doing action a given the current state s. The action which produces the highest Q-value is the action chosen by the robot. Occasionally, a random action is taken to explore new state-action sequences.

This Q-value is also used in updating the "conductance" of each cell in the just-used Q-map once a reward value, r, has been received. The following expression is calculated for each cell.

$$G'_{i,j,m,a} = \lambda(G_{i,j,m,a} + \rho(r + E(S') - G_{i,j,m,a}))$$

Where $E(S')$ is the highest Q-value promised by any of the other Q-maps in the same module for the current (post-action) state. The memory decay parameter λ, set to 0.99 for experiments, will make the Q-maps gradually forget erroneous sequences. Correct sequences will not be affected by this parameter because the decay is very weak compared to the strengthening produced over time by Q-learning. The learning rate function, ρ, controls how fast the robot alters the conductance of a cell given the probability that the cell is occupied. For the experiments we initially set $\rho = 0.5 s^p_{i,j}$.

Q-learning can also be applied to the closest cells within k cell units of the cell being updated. This helps generalize across states where objects are in similar but not identical positions. It is also another way to indirectly specify the adjacency structure of the inputs cells. We incorporate this into the Q-learning rule by changing the learning rate function. It now becomes:

$$\rho = \frac{s^p_{i,j}}{2} + \sum_{n,m:d<k} \frac{s^p_{n,m}}{2d}$$

where d is the distance from the principle cell $(s^p_{i,j})$ to each peripheral cell $(s^p_{n,m})$. Thus a peripheral cell's affect on the current cell's value is proportional to the inverse of the distance between them.

3.3 Performance

Using the same environment and methodology as before, experimental data were collected on learning runs utilizing the Q-map approach. Briefly, each learning run for the mobile robot was composed of 20 trials, each trial lasting 100 steps, for a total of 2000 steps. The goal is the same as before: to find boxes, push them once found, and "unwedge" from stalled situations. However, instead of a real robot, a simulator was used for these experiments. This made experimentation easier, but the simulator is deceptively "cleaner" than the real world. For instance, actions are always exactly invertible and there is no error in the sonar ranges returned. Also, boxes do not rotate in the simulator, they only slide.

Table 2 compares Q-maps with an extended form of OBELIX's clustering algorithm which works with 144 real valued numbers (the cell occupancies) instead of just 18 binary inputs. Clustering on this improved input shows a noticeable increase in the performance of the finder module. Finding requires the robot to detect and recognize boxes at a distance. Apparently the improved spatial representation can help overcome the normally mediocre performance on this subtask. However, Clustering does not do so well on the other two subtasks, and is actually worse than the original method. In general, the robot only needs knowledge of nearby objects to solve these subtasks. The poor performance of Clustering suggest that it gets confused by the new wealth of information and cannot effectively solve the spatial generalization problem in the limited number of steps used in this experiment. By contrast, the Q-map approach performs the individual tasks about as well as the old system – it does not seem to be confused. However, it fails to take advantage of long-range information to improve its performance on the finder subtask. This may be due to the fact that radially smoothing the scrolling occupancy grid with a Gaussian tends to discount the importance of far off objects too much.

Technique	Finder	Pusher	Unwedger
Clustering (18 bits)	55%	88%	75%
Clustering (144 cells)	71%	50%	50%
Q-maps (144 cells)	48%	84%	72%

Table 2. Average reward performance at end of learning run on the simulator.

Looking at Table 3, we see that both clustering (144 cells) and the Q-map approach perform the overall task about twice as well as the old 18 bit clustering technique. This shows that the improved input representation allows improved performance. This is likely to be due to the fact that both techniques are better able to find real boxes. This is not directly reflected in Table 2 since the finder reward is positive only when there is something aligned with the robot and it is going forward (cf. [Mahadevan and Connell 1992]). All the steps involved in rotating toward the box or approaching it from very far off are actually penalized! Q-learning tends to percolate the good ultimate value back through this chain of actions and hence maximize the *expected* reward. What is recorded in the table more closely reflects the average reward value than the average Q-value.

Technique	%
Q Learning with Clustering (18 bits)	8%
Q Learning with Clustering (144 cells)	14%
Q Learning with Q-maps (144 cells)	18%

Table 3.   Percentage of steps during learning run that the robot was actually pushing a box in the simulator.

Empirically, clustering was much slower on the 144 cell input as compared to 18 bits, and took much longer to compute an answer than the Q-map technique did. This is understandable given that clustering must check *every* cluster to see if a state belongs to that cluster and then whether this cluster needs to be merged with another. Q-maps have only a linear time dependence on the size of the input. This can become more of an issue if the input matrix grows very large. Still, assuming computation is very cheap, Q-maps still have an advantage over Clustering as shown in Table 2. The same radial smoothing heuristic that hinders Q-map's finder performance may also help it on task where nearness really is correlated to importance, such as the pusher and unwedger subtasks. Thus Q-maps would be the preferred technique for this problem given that Q-maps are generally easier to compute than Clustering and that, overall, they perform as well or better.

4   USING ACTION MODELS

One of the major problems with reinforcement learning is that most systems are set up to learn only one particular task. If some way could be found to transfer knowledge across tasks, the learning of multi-part

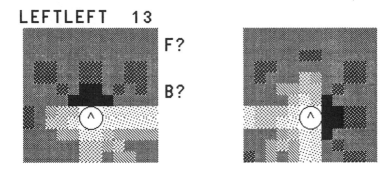

LEFTLEFT 13

F?

B?

Figure 9. An example action model showing changes to sensory bits.

tasks could be speeded up and the learning of new tasks could be substantially accelerated. This section describes such a technique based on "action models" that attempts to maximize transfer within and across tasks. These models describe how each action the robot might take will change its current sensory input. Transfer is enhanced because the action models do not contain any task specific information in them. A detailed description of this approach and the ASTERIX robot is given in [Mahadevan 1992].

Figure 9 illustrates an action model learned by a simulated robot. The probability information is depicted graphically, with the darker regions representing higher occupancy probabilities, and the lighter regions representing lower occupancy probabilities. This action model can be interpreted as follows. If the robot is currently in a state that matches the cluster on the left hand side, then taking the action LEFTLEFT (turning left by 90 degrees) will result in a state matching the cluster on the right-hand side. This action model has been formed from 13 instances.

The spatial pattern generated by the scrolling certainty grid sensor makes it easy to interpret the action model. In particular, the solid black stripe above the robot in the left-hand cluster represents an object, such as a wall. The action model can be interpreted as saying that if there is some object to the front of the robot, and the robot is bumped against it, then turning left by 90 degrees will result in the object appearing to the right of the robot in the resulting state. This model is useful when the robot is learning to avoid obstacles, or follow walls.

Note that there is an apparent circularity in learning action models from scrolling certainty grid state descriptions. The routine that translates or rotates the grid at every step (depending on the particular action taken) in some sense already has an action model built into it! Still, there are several reasons why learning action models makes sense even if this sensory temporal integration procedure is used. First, the built-in action model is an idealized one. In contrast, the learned action model can predict certain features of real environments based on experience. For instance, they can guess the new values of cells at the front boundary of the occupancy grid from the observation that objects tend to be extended in space. In addition, learned models can also account for limitations in the sensors used, such as specular reflections from sonar. Finally, the learned action model also projects three extra predicates, at least one of which (the stuck bit **S?**) cannot be predicted using the built-in action model.

4.1   Action Models

Formally, an action model a_j of an action a is a 3-tuple $\langle C_{pre}, C_{succ}, m_j \rangle$ where C_{pre} and C_{succ} are clusters representing generalized descriptions of states immediately preceding and succeeding action a, respectively. The value m_j represents the number of instances that have matched the model so far. Each *cluster* C is in turn an n tuple $\langle p_1, \ldots, p_n \rangle$, where p_i is an estimate of the probability of the ith state element being occupied. Here n is the number of elements in a state (for an $N \times N$ certainty grid, $n = N^2$).

An *action instance* $\langle s, a, t \rangle$ of an action a is a pair of states s and t that immediately preceded and succeeded (respectively) an instance of the use of action a by the agent. Deciding if an instance of an action a matches a given action model a_j requires matching the states s and t against the predecessor and successor cluster descriptions in the action model. Thus, we first describe how states are matched against clusters.

A simple criteria for deciding if a state s is an instance of a cluster C is obtained using the following "geometric" view of clusters and states. Basically, states and clusters are n-dimensional real-valued vectors, where each value lies between 0 and 1. Thus they can be interpreted as points within an n-dimensional unit hypercube. So a state can be viewed as an instance of a cluster if it is "close" enough in terms of the Euclidean distance metric on this hypercube. The problem with this criteria is that it treats all elements as equal, which is not true of most sensors. In particular, it is not true of the state description shown in Figure 8 since the three extra predicates **F?**, **B?**, and **S?** clearly have more "semantic" content that a certainty grid cell has.

To account for this disparity, we define the *weighted* Euclidean distance between a state s and a cluster c as:

$$distance(s, C) = \sqrt{\sum_{i=1}^{n} w_i (s_i - p_i)^2}$$

Here s_i and p_i are the probabilities of the ith state element and ith cluster element being occupied, respectively. The weights w_i of all the state elements is 1, except for the three extra predicates **F?**, **B?**, and **S?** which have a weight of ϵ^2, where ϵ is a parameter to the algorithm. Given this distance metric, a state s is considered an instance of a cluster C if $distance(s, C) < \epsilon$. Thus, by attaching an equal weight of ϵ to each of the three extra predicates, we are ensuring that a state and a cluster differing in any one of these predicates will not match because the overall distance between them will be at least ϵ. Finally, an action instance $\langle s, a, t \rangle$ matches an action model $a_j = \langle C_{pre}, C_{succ}, m_j \rangle$ if s matches C_{pre}, and t matches C_{succ}.

4.2   Temporal Projection

The utility of doing an action a in a state s can be defined as shown in Figure 10. The basic idea is to select the action model of a that best matches the current state, and recursively estimate the utility of the next state, where the next state is essentially the successor cluster description of the model. $R(s, a, t)$ is the reward obtained on doing action a in state s which resulted in state t. Given a way of estimating $U(s, a, l)$, where

$U(s, a, l)$:
1. If the lookahead distance $l = 0$, return 0.
2. Retrieve action model $a_j = \langle C_{pre}, C_{succ}, m_j \rangle$
   that maximizes the quantity
   $$P(s, C_{pre}) = \frac{\sqrt{n+3\epsilon^2} - distance(s, C_{pre})}{\sqrt{n+3\epsilon^2}}$$
   where $\sqrt{n + 3\epsilon^2}$ is the maximum weighted Euclidean
   distance (the n cells in the scrolling occupancy grid
   could have weight 1, and the 3 extra predicates
   could have weight ϵ each).
3. Let the predicted next state $t = C_{succ}$
4. Return $U(s, a, l) = P(s, C_{pre})R(s, a, t) +$
   $\max_{a'} \gamma U(t, a', l - 1)$
   where γ is a discount factor less than 1.

Figure 10. A limited-depth search algorithm for computing the utility of doing an action in a state.

l is the depth of the search, then the best action in any given state is simply the one which has the highest $U(s, a, l)$ value.

Figure 11 illustrates the lookahead procedure just described for a wall following robot. The original state is shown in the first row. The second row shows the possible next states. Only two actions are shown for clarity. The third row shows the possible states obtained by expanding the states in the second row. Again for clarity, only two actions are shown, one of them different from before for diversity. The actions taken are indicated above the state resulting from the action. The first number after the action indicates the quality of the match between the state and the best matching model for the action. The second number indicates the reward for the particular action instance.

In the original state, the robot is bumped (**B?**) against some object in front (**F?**) of it. The best matching action model for the FORWARD action predicts with probability 0.93 that moving forward will result in the robot being stuck (**S?**), an undesirable transition indicated by the reward of -1.0. The best sequence of moves for the robot is to turn right by 90 degrees (RIGHTRIGHT), and then to move forward along the "wall" on its left. Note that the topmost state and the rightmost state

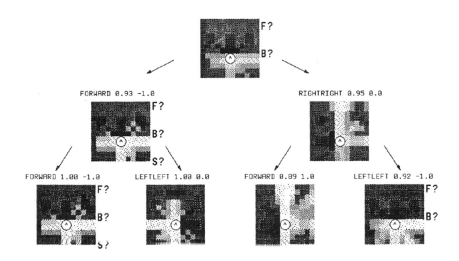

Figure 11. Action models can be used to generate a search tree of future states.

on the bottom match fairly closely, as they should since RIGHTRIGHT is the inverse of the LEFTLEFT action.

Note that the lookahead procedure in Figure 10 is an approximation to a more accurate (and intractable) lookahead procedure, which would involve exploring every possible action model of every action [Watkins 1989]. Even so, the complexity of the lookahead procedure in Figure 10 is at least $|a|^{l-1}|M|$, where $|a|$ is the number of actions, $|M|$ is the average number of action models, and l is the lookahead distance. Thus, it is exponential in the number of actions, and can get very expensive when the number of actions and the lookahead distance is very large. We have not systematically experimented with heuristics for reducing the lookahead time for two reasons. First, the number of actions in our case is relatively small (5) and we have only needed small amounts of lookahead (less than 5). Furthermore, long lookaheads are undesirable anyway as the quality of temporal projection rapidly diminishes as the number of lookahead steps is increased.

4.3   Experiments

We now describe two robot tasks that we use to illustrate transfer across
and within tasks. The specification of the box pushing task remains the
same as in Section 2. In particular, the task is decomposed into three
separate behaviors: box finding, box pushing, and unwedging.

The second task, following along a wall, is decomposed into the separate
behaviors: wall finding, and wall following. Since finding a wall must
precede following one, the module in charge of finding is given the lowest
priority and is always active. If a potential wall is found, the agent must
try to follow it, and thus the wall-following module overrides the wall
finder whenever it is active. Its applicability condition is triggered by
either the bump predicate **B?** being true, or when a "wall" predicate
W? is true. **W?** is true when when there are two contiguous cells which
have occupancy probabilities greater than 0.6 in either of two particular
areas of the scrolling grid. These special areas consist of those cells which
are in the same set of rows as the robot, and in the directly adjacent
column on either the robot's left or right side The wall follower remains
active for an additional 5 steps after its triggering condition turns off.

Knowledge of the wall-following task is specified by a reward function
for each behavior. For example, the **F?** predicate is true whenever one of
the cells in a small patch directly in front of the robot has a sufficiently
high occupancy value. If the agent is finding walls and **F?** is false, the
reward received is -1. If it goes forward and causes **F?** to come on when
before it had been off, it receives a reward of +3. Finally, if neither of
these cases holds, the default reward is 0. If the agent is instead already
following a wall, it gets a reward of +1 for going forward when **B?** is
false and **W?** is true. The reward, however, is -1 if either **B?** is true or
W? is false (i.e. the robot is bumped or has lost the wall). Otherwise
the reward is 0.

Table 4 shows the results. With the exception of wall finding and box
finding, the improvement achieved by the action models technique is as
good or better than the original Q learning/clustering technique and the
handcoded program. This is remarkable given the low level of lookahead
(3), and the more complex sensor used (144 real values instead of 18

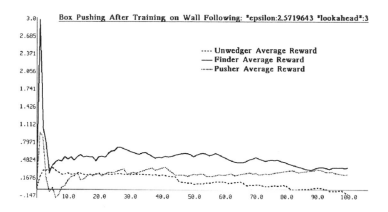

Figure 12. After learning to follow walls, the robot is able to learn to push boxes more quickly. The horizontal axis is the percentage of the trial completed while the vertical axis is average reward so far.

bits). Furthermore, a lower average reward at box finding does not necessarily translate to poorer overall task performance. In terms of overall performance, on 16% of the steps the system was actually pushing a box. This is much better than the 8-10% figure obtained in our initial study.

Behavior	Act. Models	Q/Cluster.	Hand
Box Finder	37%	55%	48%
Box Pusher	88%	88%	88%
Unwedger	88%	75%	85%
Wall Finder	24%		32%
Wall Follower	69%		61%

Table 4. Average reward at the end of a learning run on the simulator.

Another important dimension on which to evaluate the action models technique is transfer across tasks. Figure 12 shows the performance of the robot at box pushing over a run of 2000 steps *after* it was trained on the wall following task. No additional learning was allowed to take place. The curves in Figure 12 show some variation because the robot is still taking a random action 10% of the time. Notice that the curves directly start to go above zero, clearly indicating that a lot of the information gained while wall following was also useful in the box pushing task. Comparing the heights of the curves (improvements of approximately

38%, 84%, and 79% for finding, pushing, and unwedging respectively) with the numbers in Table 4 shows that the performance is close to that when the robot was directly trained on the box pushing task. Thus when one task has already been learned, learning another is easy.

5 HIGHLY STRUCTURED LEARNING

A major part of navigating within an office building is compensating for the variability of the environment. There are often people in the halls, doors open and close by random amounts, some days there are large stacks of boxes, and the trash cans are always moving around. Also, even when there are no dynamic changes, the actual fine-grained geometric details of each hallway are different. Despite these problems, TJ (a robot similar to the one shown in Figure 1) is actually able to map a collection of corridors and doorways and then rapidly navigate from one office in the building to another (average speed = 32 inches/second).

We solved the huge underlying spatial generalization problem at a high level by basically ignoring it. Instead of plotting a detailed trajectory for the robot, we depend on a behavior-based system to reliably follow a path which has been only vaguely specified. Thus we can get away with very coarse geometric maps: the robot just remembers the distance and orientations between relevant intersections. By reducing the input space so drastically, we greatly speed up the "learning" process. Furthermore, the required memorization activity only occurs when the robot reaches an intersection. Since large amounts of real time are automatically compressed into a single intervals, the temporal credit assignment problem becomes much easier as well.

5.1 Architecture

In order to exploit its collection of innate behaviors, the TJ robot uses a special control architecture called "SSS", an acronym for "servo, subsumption, symbolic" system [Connell 1992]. This architecture attempts to combine the best features of conventional servo-systems and signal processing, with multi-agent reactive controllers and state-based symbolic AI systems.

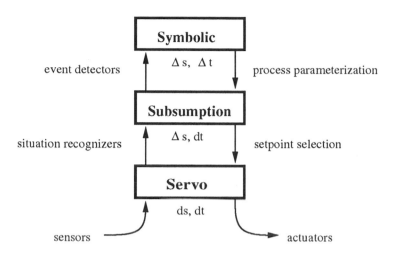

Figure 13. For navigation we graft a symbolic system on top of a subsumption controller.

The three layers in our system come from progressively quantizing first space then time. As shown in Figure 13, servo-style systems basically operate in a domain of continuous time and continuous space. That is, these systems constantly monitor the state of the world and typically represent this state as an ensemble of scalar values. Behavior-based systems also constantly check their sensors but their representations tend to be special-purpose recognizers for certain types of situations. In this way behavior-based systems discretize the possible states of the world into a small number of special task-dependent categories. Symbolic systems take this one step further and also discretize time on the basis of significant events. They commonly use terms such as "after X do Y" and "perform A until B happens". Our system creates these temporal events on the basis of changes in spatial situations. Thus, it does not make sense for us to discretize time before space. For the same reason, our system does not include a fourth layer in which space is continuous but time is discrete.

In order to use these three fairly different technologies we must design effective interfaces between them. The first interface is the command transformation between the behavior-based layer and the underlying servos. This is accomplished by letting the subsumption-style behaviors adjust the setpoints of servo-loops, such as the wheel speed controller, to one of a few values. All relevant PID calculations and trapezoidal profile generation are then performed transparently by the underlying servo system.

The sensory interface from a signal-processing front-end to the subsumption controller is a little more involved. A productive way to view this interpretation process is in the context of "matched filters" [Wehner 1987, Connell 1991]. The idea here is that, for a particular task, certain classes of sensory states are equivalent since they call for the same motor response by the robot. There are typically some key features that, for the limited range of experiences the robot is likely to encounter, adequately discriminate the relevant situations from all others. Such "matched filter" recognizers are the mechanism by which spatial parsing occurs.

The symbolic system grafted on top of the subsumption layer must also be able to receive information and generate commands. It terms of control, we have given the symbolic system the ability to selectively turn each behavior on or off, and to parameterize certain modules. These event-like commands are "latched" and continue to remain in effect without requiring constant renewal by the symbolic system.

On the other end, information is transferred from the behavior-based layer to the symbolic layer by a mechanism that looks for the first instant in which various situation recognizers are all valid. For example, when the robot notices that it has not been unable to make much progress toward its goal recently, a "path-blocked" event is signalled to the symbolic layer. However, many times the symbolic system can not handle the real-time demands imposed by such events. Thus, to help decouple things we have added a structure called the "contingency table". This table allows the symbolic system to *pre-compile* what actions to take when certain events occur, much as baseball outfielders yell to each other "the play is to second" before a pitch. The entries in this table reflect what the symbolic system expects to occur and each entry essentially embodies a one-step plan for coping with the actual outcome.

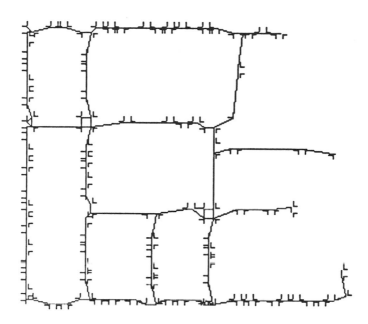

Figure 14. A manually surveyed map of one side of our building. The robot only needs to know the rough lengths of the corridors and their angles at the intersections.

5.2   Maps and Planning

The strategic part of navigation - where to go next - is handled by the symbolic layer. To provide this information, our symbolic system maintains a coarse geometric map of the robot's world. This map consists of a number of landmarks, each with a type annotation, and a number of paths between them, each with an associated length.

The landmarks used in the map are the sudden appearance and disappearance of side walls. These locations can be detected using the long range IR proximity detectors on each side of the robot. Normally, in a corridor the robot would continuously perceive both walls. When it gets to an intersection, suddenly there will be no wall within range of one or both of these sensors. Similarly, when the robot is cruising down a corridor and passes an office, the IR beam will enter the room far enough so that no return is detected.

The path lengths can be easily measured by odometry. However, to successfully build a map the robot must also be able to determine when it has arrived back in a place it has been before. The difficulty is that, using standard robot sensory systems, an individual office or particular intersection is not distinguishable from any other. One approach is to integrate the odometric readings to determine the absolute position of the robot. However, it is well known that over long distances such measurements can drift quite severely due to differing surface traction and non-planar areas.

We solve the loop problem by exploiting the geometry of the environment. In most office buildings all corridors are more or less straight and meet at right angles. Therefore we measure the length of each path segment and treat this as a straight line. Similarly, when the robot switches from one path to another we force the turn to be a multiple of 90 degrees. This is essentially an odometric representation which is recalibrated in both heading and travel distance at each intersection. In this way we maintain a coarse (x, y) position estimate of the robot which can be compared to the stored coordinates of relevant places.

The tactical part of navigation – how to actually traverse some segment of a path – is handled by the subsumption layer. Once a map has been created, either manually or by the robot, an efficient route can be plotted by a spreading activation algorithm [Mataric 1992] or some other method. The symbolic system then enables whatever collection of subsumption modules it deems appropriate for the first segment of this path and parameterizes the operation of the relevant modules in certain ways. It might, say, turn on obstacle avoidance and corridor following, set the travel distance to 30 feet, and activate the intersection detector. In general, the symbolic system does not need to constantly fiddle with the subsumption layer; it only has to reconfigure the layer when specific events occur. For instance, as soon as the robot reaches the end of the current path segment the robot's course needs to be changed to put it on the next segment. The contingency table allows such alterations to be accomplished swiftly so that the robot does not accidentally do things such as overshoot the intersection.

5.3 Performance

Figure 15 shows a map produced by the real robot. In general, we could create a map by letting the robot wander around its environment performing some sort of tree search on the network of corridors. For this example we instead provided the robot with a rough path to follow of the form:

$$((travel_1, turn_1)(travel_2, turn_2)\ldots).$$

Travel was specified as an integral number of inches and turns were specified as an integral number of degrees. The top half of Figure 15 shows the path the robot took according to its odometry. The circles with small projections indicate where the robot observed openings in the directions indicated. The circles are drawn the same size as the robot (12 inches diameter) to provide scale. Notice that neither of the loops appears to be closed. Based on this information it is questionable whether the corridor found in the middle of the map is the same one which the robot later traversed.

The symbolic map is shown in the lower half of Figure 15. Nodes are denoted iconically as corners – two short lines indicating the direction of the opening and the direction of free travel. This symmetry reflects the fact that the robot is likely to perceive the same corner when coming out through the marked aperture.

Notice that the symbolic map correctly matches the important intersections to each other. When a new opening is detected the robot compares it with all the other nodes in the map that have similar opening directions. If there is a match which is no more than 6 feet from the robot's current position in either x or y, the new opening is considered to be a sighting of the old node. In this case we average the positions of the robot and the node and move both to this new location. This merging operation is why the corridors do not look perfectly straight in the symbolic map. However, when the robot is instructed to follow the same path a second time and update the map, the changes are minimal.

Although the odometric trace shows many local maneuvering steps, these actions never percolate up to the symbolic system. The mapper only wakes up at the intersections or if something goes wrong. Still, despite

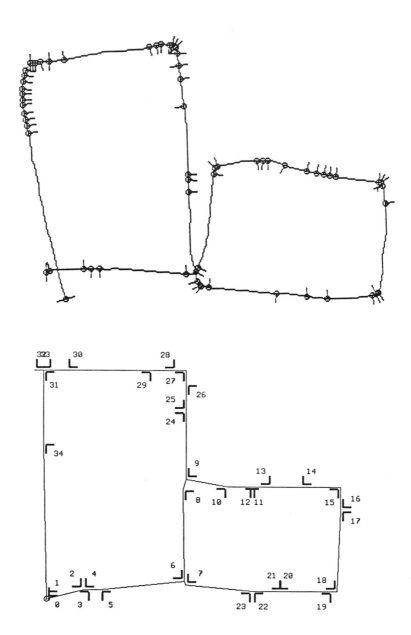

Figure 15. The path traversed according to odometry (top) and the symbolic map learned from this data (bottom).

being oblivious to most of the details around it, the symbolic system receives enough information to build up a simple representation of the world which is sufficient to allow the robot to navigate purposively. By constraining the robot to pay attention to only a few special parameters of its environment, we allow it to *quickly* learn what it needs to know.

6 SUMMARY

In general, we believe *tabula rasa* learning is unlikely to work on real robots. The space of possible control functions is extremely large, and real robots can only be run for a limited time. However, there are many things the robot can be told that will make its learning task significantly easier. The key is finding out what those things are and then devising ways of incorporating them into the robot. In the box-pushing example we gave the robot a rough sketch of a procedure, and let it fill in the details. When we increased the size of the input space by using scrolling occupancy grids, we did not treat each of the cells as an independent value. We told the robot about the neighborhood relations of each cell and gave it a hint that far away cells were not as important as nearby ones. In the action-model example we implicitly told the robot that the world reacts the same way no matter what the robot's goal happens to be.

As an extreme example, in the mapping experiment we drastically whittled down the size of the spatial and temporal credit assignment problems by exploiting the competence of the corridor following behavior as well as the spatial regularity of the environment. We believe that much of the "during lifetime" learning that is useful to robots may actually be of this form. The designer builds in as much as he knows about the task, environment, and sensors then lets the robot only make minor adaptations to changing circumstances and determine those few facts which could not be known a priori.

Acknowledgements

This research was funded by the IBM Corporation and performed at the T.J. Watson Research Laboratory in Hawthorne, New York.

6

THE SEMANTIC HIERARCHY
IN ROBOT LEARNING

Benjamin Kuipers
Richard Froom
Wan-Yik Lee
David Pierce

Artificial Intelligence Laboratory,
Department of Computer Sciences,
University of Texas at Austin, Austin, TX 78712

ABSTRACT

We have been exploring an approach to robot learning based on a hierarchy of types of knowledge of the robot's senses, actions, and spatial environment. This approach grew out of a computational model of the human cognitive map that exploited the distinction between procedural, topological, and metrical knowledge of large-scale space. More recently, the semantic hierarchy approach has been extended to continuous sensorimotor interaction with a continuous environment, demonstrating the fundamental role of identification of *distinctive places* in robot spatial learning. In this paper, we describe three directions of current research. First, we are scaling up our exploration and map-learning methods from simulated to physical robots. Second, we are developing methods for a *tabula rasa* robot to explore and learn the properties of an initially uninterpreted sensorimotor system to the point where it can reach the control level of the spatial semantic hierarchy, and hence build a cognitive map. Third, we are developing learning methods for incrementally moving from low-speed, friction-dominated motion to high-speed, momentum-dominated motion. Achievement of these goals will bring us closer to a comprehensive computational model of the representation, learning, and use of knowledge about space.

1 INTRODUCTION

We have been exploring an approach to robot learning based on a hierarchy of types of knowledge of the robot's senses, actions, and spatial environment. This approach grew out of a computational model of the human cognitive map that exploited the distinction between procedural, topological, and metrical knowledge of large-scale space [Kuipers 1978, Kuipers 1979b, Kuipers 1979a, Kuipers 1983a, Kuipers 1983b]. More recently, Kuipers and Byun [Kuipers and Byun 1988, Kuipers and Byun 1991] extended this semantic hierarchy approach to continuous sensorimotor interaction with a continuous environment, demonstrating the fundamental role of identification of *distinctive places* in robot spatial learning. Our current research extends the semantic hierarchy framework in three directions.

- We are testing the hypothesis that the semantic hierarchy approach will scale up naturally from simulated to physical robots. In fact, we expect that it will significantly simplify the robot's sensorimotor interaction with the world.

- We are developing methods whereby a *tabula rasa* robot can explore and learn the properties of an initially uninterpreted sensorimotor system to the point where it can define and execute control laws, identify distinctive places and paths, and hence reach the first level of the spatial semantic hierarchy.

- We are demonstrating how the semantic hierarchy and the learned topological and metrical cognitive map support learning of motion control laws, leading incrementally from low-speed, friction-dominated motion to high-speed, momentum-dominated motion.

We describe progress toward these goals in the sections below. If these goals can be achieved, we will have formulated a comprehensive computational model of the representation, learning, and use of a substantial body of knowledge about space and action. In addition to the intrinsic value of this knowledge, the semantic hierarchy approach should be useful in modeling other domains.

2   THE COGNITIVE MAP AND THE SEMANTIC HIERARCHY

Spatial knowledge is central to the ability of a human or robot to function in a physical environment. Spatial knowledge plays a foundational role in common-sense knowledge generally. Spatial knowledge is also a particular challenge to the Physical Symbol System Hypothesis [Newell and Simon 1976] since it requires effective symbolic representations of continuous reality. In this paper, we describe the foundations of our approach to representing and learning spatial knowledge, along with the results of recent past and current research.

The *cognitive map* is the body of knowledge a human or robot has about its large-scale environment. (An environment is *large-scale* if its spatial structure is at a significantly larger scale than the sensory horizon of the observer.) Observations of human spatial reasoning skills and the characteristic stages of child development provide clues as to how the cognitive map is represented and learned [Lynch 1960, Piaget and Inhelder 1967, Moore and Golledge 1976]. Computational constraints provide additional clues [Kuipers 1979b, Kuipers 1983a]. Based on these clues and constraints, we have developed an extensive theory of the human and robotic cognitive map [Kuipers 1977, Kuipers 1978, Kuipers 1979a, Kuipers 1982, Kuipers 1983b, Kuipers and Levitt 1988, Kuipers and Byun 1988, Kuipers and Byun 1991].

Our theory of the cognitive map is built around two basic insights due to many scientists, most notably Jean Piaget. The first is that a *topological* description of the environment is central to the cognitive map, and is logically prior to the metrical description. The second is that the spatial representation is grounded in the sensorimotor interaction between the agent and the environment. Our concept of *distinctive place* provides a critical link between sensorimotor interaction and the topological map.

This leads to a three-level spatial representation called the *spatial semantic hierarchy* (SSH):

$$[sensorimotor \leftrightarrow control] \rightarrow topology \rightarrow geometry.$$

Each level of the hierarchy (see Figure 1) has its own ontology for describing the world, and supports its own set of inference and problem-solving methods. The hierarchy fits together because each level provides the properties that the following level depends upon [Kuipers and Levitt 1988].

At the *control level* of the hierarchy, the ontology is an egocentric sensorimotor one, consisting of sensory and motor primitives, and composite features and control strategies built out of the primitives. In particular, the ontology does not include fixed objects or places in an external environment. A distinctive place is defined as the local maximum found by a hill-climbing control strategy, moving the robot to climb the gradient of a selected sensory feature, or *distinctiveness measure*. Distinctive paths are defined similarly. The critical task at this level is to decide which sensory feature to use as a distinctiveness measure to drive the current control strategy. These feedback-guided control strategies provide robust performance, mitigating the effects of sensor and motor uncertainty. In particular, for motion between distinctive places, cumulative position error is eliminated.

At the *topological level* of the hierarchy, the ontology consists of the places and paths defined by the control level, with relations connecting them into a network. (This can be extended to include regions and containment relations [Kuipers and Levitt 1988].) At the topological level, the critical task is to link places and paths into a network, deciding whether the current place is new or a familiar place revisited (cf. [Dudek *et al.* 1991]). The network can be used to guide exploration of new environments, to solve new route-finding problems, and to plan experiments to disambiguate places with indistinguishable sensory images. The network representation means that adequate navigation is not dependent on the accuracy, or even the existence, of metrical knowledge of the environment.

At the *geometrical level* of the hierarchy, the ontology of fixed places and paths in an external environment is extended to include metrical properties such as distance, direction, shape, etc. Only at this level are sensory features considered to be transduced properties of the environment. Initially, geometrical features are extracted from sensory input, and represented as annotations on the places and paths of the

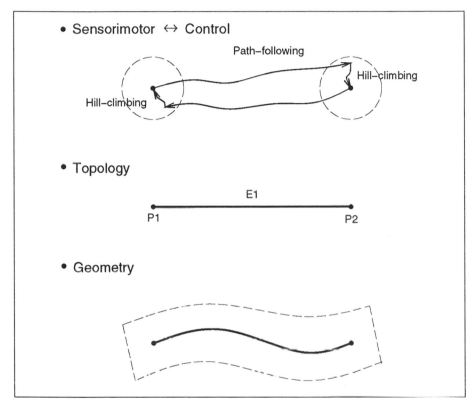

Figure 1. The Spatial Semantic Hierarchy representation allows each level to establish the assumptions required by later levels.

- **Control.** When traveling between distinctive places, cumulative error is eliminated by alternating path-following with hill-climbing control strategies.

- **Topology.** Eliminating cumulative error permits the abstraction from a continuous world to a topological network description.

- **Geometry.** Place and path descriptions are annotated with geometrical information, e.g., generalized cylinder or occupancy grid descriptions.

Reprinted from [Kuipers and Byun 1991].

topological network. For example, each place could be annotated with an occupancy grid representation of its shape, while each path would be annotated with an associated generalized cylinder model. Since the topological structure of the environment has been provided by the previous level, the local metrical descriptions can be relaxed into a global metrical model of the environment as a whole.

The levels of the hierarchy support each other and depend on each other in intricate ways. The control-level definition of places and paths provides primitive elements for the topological description, which in turn supports navigation while the more expensive sensor fusion methods accumulate metrical information. When metrical information is available, it can be used to optimize travel plans or to help disambiguate indistinguishable places, but navigation and exploration remain possible even without metrical information. At any given place, the dependencies described by the spatial semantic hierarchy constrain what knowledge can be represented in what order. Nonetheless, in the cognitive map as a whole, different places and regions can be represented at different levels.

Our approach contrasts with more expensive and fragile traditional methods, which attempt to acquire geometrical information directly from sensor input and then derive topology by parsing the geometrical description (e.g., [Chatila and Laumond 1985, Moravec and Elfes 1985]). From our perspective, the traditional approach places geometrical sensor interpretation (the most expensive and error-prone step) on the critical path prior to creation of the topological map, which is both intrinsically easier to build and is helpful in building the metrical map.

Our spatial semantic hierarchy is consistent with Brooks' [Brooks 1986] subsumption architecture: the control level corresponds roughly with Level 2, 'Explore', and the topology and geometry levels correspond with Level 3, 'Build maps.' Most implementations of the subsumption architecture (e.g., [Brooks 1990a, Gat and Miller 1990]) are hierarchies of simple finite-state controllers that reject map-like representations in favor of reflex-driven behavior. In contrast with the currently popular rejection of map-like representations [Brooks 1991a], we believe that the spatial semantic hierarchy provides a principled relationship between the reflex-like control level and the map-like topological and metrical levels of spatial knowledge.

2.1 Exploration and Mapping Experiments

Kuipers and Byun [Kuipers and Byun 1988, Kuipers and Byun 1991] demonstrated the SSH approach on a simulated robot called NX, showing that it could learn an accurate topological and metrical map of a complex two-dimensional environment in spite of significant sensory and motor errors. The environment is assumed to be fixed, except for the position of the robot. NX simulates a robot with a ring of distance sensors, an absolute compass, and tractor-type chain drive, with sensor error models suggested by the properties of sonar range-finders [Drumheller 1987, Flynn 1985, Walter 1987].

NX was tested on a large number of two-dimensional environments. Its performance is illustrated by the following figures (from [Kuipers and Byun 1991]).

- **Exploration and Control.** Figure 2a shows the tracks followed by NX while exploring its environment. All distance readings are subject to 10% random error, and heavily blocked walls yield specular reflection errors. Black dots represent locations at which it identified a distinctive place. Clusters of dots represent the positions at which the robot identified the *same* distinctive place on different occasions. Loops and repeated path traversals represent exploration strategies to resolve various ambiguities.

- **Topology.** Figure 3 shows a fragment of the topological map, corresponding to the left side of Figure 2a. Places and paths are labeled with their associated control strategies.

- **Geometry.** Figure 2b shows the metrical map obtained by relaxing local generalized-cylinder descriptions of the paths and places into a single frame of reference. The metrical map is a reasonably accurate match for the actual environment in Figure 2a.

In addition, the SSH framework helps us clarify the sources of certain mapping problems, and to determine how knowledge from different levels in the hierarchy can be combined to solve them (cf. Figure 4).

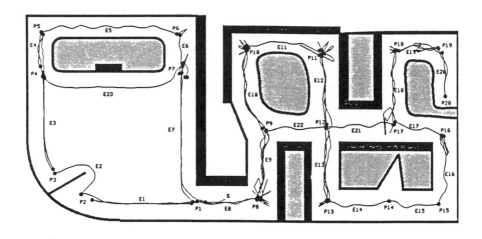

(a)

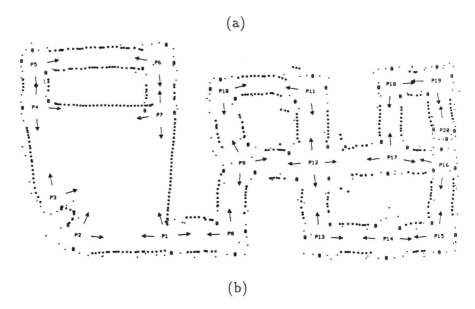

(b)

Figure 2.   NX explores and maps a complex environment.

(a) Different physical points and routes can correspond to the same distinctive places and paths.

(b) The metrical annotations on the topological map can be relaxed into a global geometrical map of the environment.

Reprinted from [Kuipers and Byun 1991].

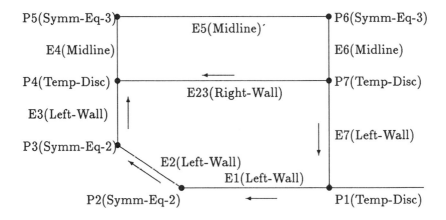

Figure 3. A fragment of the topological map. Annotations are names of distinctiveness measures or control strategies defining place and paths, respectively. *Reprinted from [Kuipers and Byun 1991].*

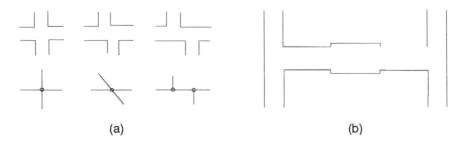

Figure 4. Topological ambiguity problems.

(a) **The bifurcation problem**: very similar sensory images give rise to different topological descriptions near a bifurcation point in the behavior of the interpretation algorithm.

(b) **The NYSINYD problem**: *"Now You See It; Now You Don't."*
A minor feature may be viewed as a distinctive place on one occasion, but overlooked while traversing a path on another.

Metrical information, when available, can help disambiguate both problems.

3   FROM SIMULATED ROBOT TO PHYSICAL ROBOTS

To evaluate the viability of the spatial semantic approach in a complex natural world, we are implementing it on two physical mobile robots. To this end, work has been done in several directions.

Several of the implementation issues are most conveniently studied on a simulator, while others necessitate experimentation with physical robots. We have developed a new robot simulator, SIMRX [1991], and prepared two physical robots, Spot and Rover, to facilitate the experimentation and evaluation of extensions to the SSH approach. Since the SSH approach is only weakly dependent on the particular sensorimotor system of a robot, a principled method of implementing the SSH approach on physical robots based on a minimal set of *abstract mapping senses* is developed. The following sections briefly describe the work done to date.

3.1   Evaluation Platforms

Each level of the SSH has its own spatial representation and inference methods. This modularity allows us to use different platforms efficiently in evaluating possible extensions to the SSH approach. SIMRX [Lee 1991], a successor of NX, complements the physical robots as an evaluation platform (Figure 5). It also allows us to study and efficiently test certain implementation and conceptual issues that are relatively independent of the way the sensor and effector systems are simulated. Particularly, it allows rapid debugging of semantic issues and prototyping of extensions to the SSH approach.

However, the simulator does have some limitations. In the simulator, the sensorimotor interaction of the robot with its surrounding is based on simplified models. The range-sensor model returns the distance to the nearest object along a ray rather than within a cone of finite angle. A simple model of the specularity problem is used. Noise of uniform distribution is introduced to the effector system to simulate inherent effector inaccuracies and errors due to slippage. Thus, in confronting certain issues, particularly those that involve sensorimotor interaction with the

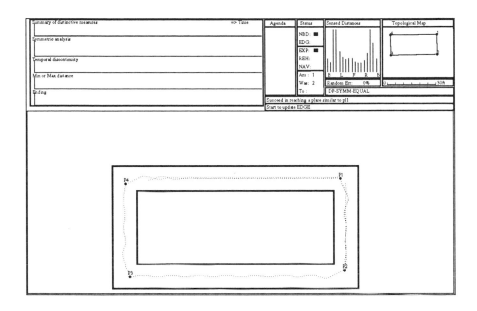

Figure 5. A typical SIMRX display showing the result of a low speed exploration of a simple environment. The learned topological map is shown in the top right box.

world, the simulator is inappropriate and physical robots are necessary for thorough evaluation of our approach. Therefore, two physical robots, Spot and Rover, are being used to help in evaluating the SSH approach.

Spot is a Real World Interface^{TM} (RWI) robot with a one foot diameter three-wheeled base and an enclosure for housing twelve sonar transducers, a power supply, a backplane, an interface board, a sonar controller board and a 68000 microcomputer. The base has its own microcontroller which accepts velocity, acceleration, translational position and rotational position commands. All three wheels and the enclosure always face the same direction. Its twelve sonar sensors are arranged uniformly in a ring at a height of approximately one foot from the ground.

Rover is a home-brewed robot with a base with two motorized wheels which are controlled based on differential velocity, and a sonar system mounted on top of the base. The base has a dedicated microcontroller to allow it to accept various motion commands through a serial port. Encoders are mounted on the motors to provide motion feedback. The sonar system consists of sixteen Polaroid sonar transducers mounted uniformly in a ring, and a microcontroller together with two Polaroid sonar drivers to accept various sensing commands and to control the operation of the sensing system. Rover is also equipped with a flux-gate compass.

The physical robots, Spot and Rover, with two different sensorimotor systems, provide two different physical platforms on which the SSH approach can be tested. In the process of implementing the SSH approach on different physical robots, we hope to develop principled engineering methods for implementing the SSH approach on a given robot with given sensorimotor capabilities. To that end, we are developing a formal specification for the sensorimotor interface of a robot to the SSH modules for exploration, mapping and navigation as described in the following section.

3.2   Abstract Mapping Senses

Since the SSH approach to exploration and map learning is relatively independent of the sensorimotor system of the robot, we are defining a set of *abstract mapping senses* to capture the minimal requirements that the SSH control level places on the sensory capabilities of the robot. The set of abstract mapping senses we define are in a sense, the "weakest preconditions" of the SSH mapping strategy. We believe there are only a few reasonable abstract mapping senses. Most of the physically realized senses described in the literature can be captured by the following abstract mapping senses:

- **Continuous-range sense:** delivers the distance to the nearest obstacle in all directions surrounding the robot. Sonar used by some mobile robots [Elfes 1987, Crowley 1985, Leonard *et al.* 1990] for mapping purposes are examples of the realization of this abstract sense.

- **Landmark-vector sense:** delivers the vectors from the robot to a set of surrounding landmarks as illustrated in Figure 6. Active beacons provide a direct realization of this sense as in [Kadonoff *et al.* 1986]. Vision can also be used to implement this abstract sense by focusing on identifying strong and stable features in the environment as demonstrated by Levitt and Lawton [Levitt and Lawton 1990] with their viewframes construct, Crowley [Crowley *et al.* 1991], Engel [Engel 1989] and Wells [Wells III 1989].

- **Open-space sense:** determines the directions in which there is sufficient open space for travel from the current position of the robot. This abstract sense can be considered a degenerate case of the continuous-range sense.

- **Place-recognition sense:** determines if the currently sensed sensory pattern or signature of the current location matches any of the stored sensory patterns or signatures of known places.

- **Continuous-feature sense:** delivers a continuous feature in all directions surrounding the robot. Examples of such features are smell, surface texture, and ground elevation.

- **Proprioceptive sense:** ("internal-sensing") encapsulates the use of internal sensors that respond to the internal state of the robot, such as odometers.

Realization of the Abstract Mapping Senses

An abstract mapping sense can be implemented using a fusion of information from two or more physical sensory systems. With the definition of abstract mapping senses, sensory fusion can be performed with a sharper focus and explicit purpose.

Any given robot can have some or all of these abstract mapping senses, realized to different degrees of perfection depending on the physical sensors available to the robot. More than one type of sensor can be used to implement an abstract mapping sense as illustrated in Figure 7.

Benefits of Abstract Mapping Senses

By defining the abstract mapping senses to form the basis for the SSH approach, we provide a generic SSH strategy defined on the abstract senses, allowing us to decompose the SSH implementation task into two independent subtasks:

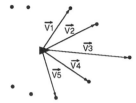

Figure 6. The landmark-vector sense provides vectors (direction and distance) from the robot to some set of the surrounding landmarks. In this example, the landmark-vector sense provides the vectors, $\vec{v}_1, \vec{v}_2, \vec{v}_3, \vec{v}_4$, and $\vec{v}_5$ to five of the landmarks, each of which is defined with respect to a local coordinate system centered at the robot.

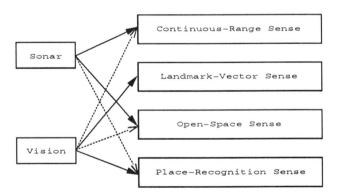

Figure 7. A robot with a sonar ring and a vision system can realize the continuous-range, landmark-vector, open-space, and place-recognition abstract mapping senses. The continuous-range sense can be realized by a sensor fusion of information from both the sonar ring and the vision system. The dashed lines are used to show a less perfect realization. The vision system can provide richer descriptions of the environment than the sonar ring and hence it realizes the place-recognition sense better than the sonar ring. The vision system can also realize the landmark-vector sense by focusing on identifying strong and stable features in the environment. The open-space sense can easily and naturally be implemented by the sonar ring.

- (re)definition of the generic SSH mapping based on abstract mapping senses, and

- implementations of the abstract mapping senses using the available sensors of the physical robot.

We hope such a decomposition can ease the implementation of the SSH approach on a given physical robot. The generic SSH mapping should require little rewriting with most effort devoted to implementing the necessary abstract mapping senses according to their specifications using the available sensors.

4 FROM *TABULA RASA* TO COGNITIVE MAPPING

In most of our work, and in virtually all other research on robot exploration, knowledge of the properties of the robot's sensors and effectors, and of their relation with the properties of the environment, is provided by the researcher and embedded in the structure of the learning system. However, this only begs the question of how this knowledge is acquired. To what extent must it be built in (i.e., innate), and to what extent can it be acquired from experience?

We conjecture that it is possible for a robot to experiment with an initially uninterpreted sensorimotor apparatus in order to learn the prerequisites for the spatial semantic hierarchy. We use the term, "critter," for such a robot.[1]

The topological and metrical levels of the spatial semantic hierarchy do not depend heavily on the nature of the sensorimotor apparatus. The control level serves as the interface between the sensorimotor system and the higher levels. The focus of *tabula rasa* map learning is therefore on developing the control level without *a priori* knowledge of the nature of the sensorimotor apparatus. The critter learns the control level of the

---

[1] The term is due to Ron Rivest. See [Kuipers 1985] for a discussion of the original formulation of the critter problem.

spatial semantic hierarchy by learning control strategies for moving to distinctive places and along distinctive paths.

4.1 Building an Ontology for the Sensorimotor System

We are applying the semantic hierarchy approach to the problem of *tabula rasa* learning just as we did to the problem of spatial learning. The critter learns local control strategies for place finding and path following through a sequence of steps. At each step, the critter acquires objects and relations that serve as the foundation for the next step.

1. **Diagnosis of structure of the sensory system.** The critter learns the relative positions of the sensors in its sensory system in order to do motion detection.

2. **Diagnosis of primitive actions.** The critter uses motion detection to identify a set of primitive actions capturing the capabilities of its motor apparatus.

3. **Identification of state variables.** The critter transforms the primitive actions into a commutative set which may be integrated with respect to time to obtain an n-dimensional state vector.

4. **Definition of local control strategies.** The critter uses constraints on sensory features to define distinctiveness measures for places and paths. Places require n constraints; paths, $n-1$. Gradient ascent (now possible since the state variables are known) is used to maximize distinctiveness measures and thus move to places and paths.

In this section, we discuss methods for accomplishing these steps for the NX robot. We have successfully demonstrated the first two steps on simulated robots with a variety of sensorimotor systems including that used by NX [Pierce 1991b]. We are currently working on automating the third and fourth steps.

Diagnosis of Sensory Structure

The first step is to group the sensors into related classes and deduce the structure (physical layout) of each class. For example, NX has a compass and a ring of range sensors. If the relative positions of the range sensors are known, then motion detection is possible since spatial as well as temporal derivatives of sensor values can be computed.

The structure of an array of sensors is reflected in the correlations among sensors. The fact that the world is approximately continuous means that sensors near each other in an array will often have similar values. The structure of the array can be captured by mapping the sensors onto a plane such that distances in the plane are proportional to dissimilarities between sensors (Figure 8b). This is accomplished using metric scaling and a simple relaxation algorithm. The dissimilarity $d(i, j)$ between two sensors is computed by summing instantaneous differences in sensor values over time: $d(i, j) = \sum_t |y_i(t) - y_j(t)|$ where y_i is the ith component of the sense vector.

The dissimilarity matrix $d(i, j)$ is used both to group sensors into subgroups of related sensors (e.g., compass vs. range sensors vs. photoreceptors) and to map each group's sensors onto the two-dimensional plane.

Diagnosis of Primitive Actions

The critter controls its motor apparatus with a real-valued action vector. It initially knows nothing about the effects of its actions or even the number of degrees of freedom of its motor apparatus. (It does not assume that the number of components of the control vector is the same as the number of degrees of freedom.) Here we show how the critter can determine the number of degrees of freedom of its motor apparatus and a set of *primitive actions* for producing motion in each degree of freedom.

For example, NX can turn in both directions and can move forward and backward. The capabilities of its motor apparatus are captured by the primitive actions *turn-left* and *turn-right* for one degree of freedom

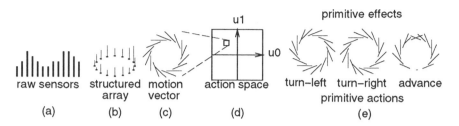

Figure 8. The diagnosis of primitive actions. Given an unordered, uninterpreted stream of sensors (a), the critter computes an intersensor dissimilarity matrix and uses it to group the range sensors together and reconstruct the sensory system's structure (here, a ring) (b), define motion detectors (c), characterize actions in terms of average motion vectors (d), and, finally, characterize this space of motion vectors using principal component analysis to determine a primitive set of actions for motion in each degree of freedom (e). The robot in this example, unlike NX, is able to move forward but not backward and thus has only one primitive action for its second degree of freedom.

and *move-forward* and *move-backward* for the second. All of the robot's actions are combinations of these primitives.

The following steps are used by the critter in diagnosing primitive actions (see Figure 8c-e, also Figure 9):

1. Define local-motion detectors using the two-dimensional map that characterizes the structure of the sensory apparatus.

2. Use these local-motion detectors to learn the *average motion vector field* (amvf) for the actions within each "box" in a discretization of action space. These characterize the effects of the actions in terms of the sensory system.

3. Analyze the set of amvf's to produce a set of qualitatively distinct amvf's (*primitive effects*) and associated primitive actions.

The third step uses principal component analysis (see [Krzanowski 1988] for a clear treatment of the subject) to produce a basis set of orthogonal eigenvectors. The effects of any action may be approximated as a linear combination of them. We interpret these eigenvectors as primitive action effects. The primitive actions are identified as the actions whose amvf's

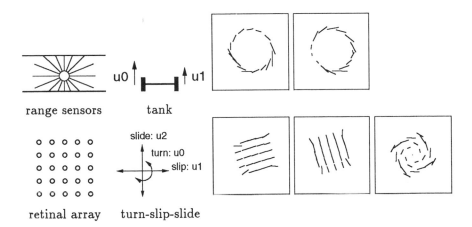

range sensors   tank

retinal array   turn-slip-slide

Figure 9 An analysis of primitive actions for two sensorimotor systems. The first is the one used by NX. The critter can move both forward and backward as well as turn. On the right are shown the two principal eigenvectors (primitive action effects) for the tank. These correspond to rotating and advancing, respectively. The second sensorimotor system includes a retinal array of photoreceptors looking down onto a picture. The critter is able to move forward and backward, left and right, and can rotate in either direction. On the right are shown the three principal eigenvectors. These correspond to sliding, slipping, and rotating, respectively.

are most collinear or antilinear with the eigenvectors. This experiment has been run with several different sensorimotor systems, two of which are illustrated in Figure 9.

Identification of State Variables

Figure 10 gives a state-space description of the critter and its environment. We are exploring methods for *tabula rasa* learning of a complete and nonredundant set of state variables which will approximate the environment's unknown state vector, x. The method we describe here involves transforming the set of primitive actions into a set of actions that capture the capabilities of the robot and are commutative. Such actions may be integrated with respect to time to produce state variables.

Consider a robot in a two-dimensional environment with cartesian coordinates x_1 and x_2 and discrete actions *right*, *left*, *up*, and *down*. The first pair of actions move the robot along the first coordinate; the second

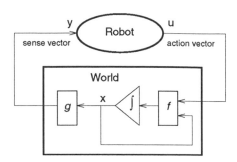

Figure 10. The *state-space description* of the robot's environment showing the relationships among actions, senses, and the state of the environment. The environment includes the sensorimotor apparatus of the robot.

pair of actions move the robot along the second coordinate. The actions are commutative. For example, moving right then up is equivalent to moving up then right (assuming the robot does not hit an obstruction). Because of commutativity, the effect of a sequence of actions does not depend on the particular order of the actions but only on the number of times that each action was taken.

If we choose the state at a particular time to be the zero state vector, then futures states may be specified by the pair $(n1, n2)$ where $n1$ is the number of *right* actions taken minus the number of *left* actions taken and $n2$ is defined analogously. These are valid state variables because the effects of any sequence of the four commutative actions is completely characterized by these numbers.

The case of discrete actions may be generalized to the case of real-valued action vectors. If the actions are commutative, then their components may be integrated with respect to time in order to obtain state variables.

As an example of noncommutative actions, consider a robot such as NX with primitive action vector $\mathbf{u} = (u_t, u_a)$ (learned by the techniques of the previous section) that can turn and advance. Its actions are not commutative: turning 90 degrees, then advancing does not lead to the same state as advancing, then turning 90 degrees. Integrating the turn-and-advance action vector with respect to time will not produce a state vector.

For such a robot, a set of commutative actions needs to be found. We have demonstrated by hand (unpublished work) how a robot with NX's sensorimotor apparatus can find a set of commutative actions and are trying to find a more general solution. A set of commutative actions may be defined in terms of the turning and advancing primitive actions using a transformation from $\mathbf{u} = (u_t, u_a)$ to $\mathbf{u}^* = (u_\theta, u_{x_1}, u_{x_2})$:

$$u_\theta = u_t \quad u_{x_1} = u_a \cos \theta \quad u_{x_2} = u_a \sin \theta$$

where θ is the state variable obtained by integrating the action component u_θ. Since $\mathbf{u}^*$ is an action vector with commutative components, $\mathbf{x}^*(t) = \int_0^t \mathbf{u}^*(\tau) \, d\tau$ may be used as a representation of the state of the environment at time t. We assume here, in contrast to Section 5, that the robot has no momentum and that velocities are not part of the state vector.

The state vector is obtained by integration of a transformed action vector and is therefore prone to cumulative error. This is not a problem since we do not use the state variables themselves when defining local control strategies; rather, we use the derivatives of features with respect to the state variables.

Definition of Local Control Strategies

Distinctiveness measures are the basis for defining the topological primitives of places and paths and may be defined in terms of constraints on sensory features. The number of constraints depends on the dimensionality of the state space. For an n-dimensional state space, n constraints are needed to define a place and $n - 1$ constraints are needed to define a path.

To illustrate, NX uses distances to nearby objects in the definition of many of its distinctive places. The distinctive place at a three-way intersection is defined as the state satisfying the constraints $y_1 = y_2$ and $y_2 = y_3$ where y_i is the distance to the ith nearby object.

The distinctiveness measure itself is a hill-climbing function that is maximized when the constraints are satisfied. A local control strategy for motion to the distinctive place is defined using gradient ascent on the distinctiveness measure. A local control strategy for motion along a path is defined using motion through state space along the one-dimensional locus satisfying the constraints that define the path.

We are currently developing a language of derived features (functions defined on the raw sense vector) guided by our intuitions about abstract mapping senses. The critter will search this language for features to use in defining distinctiveness measures for places and paths. In an earlier paper [Pierce and Kuipers 1991], we introduced techniques for defining features suitable for hill-climbing.

5   FROM LOW-SPEED TO HIGH-SPEED MOTION

Another assumption often built into robot exploration experiments is that the robot moves only when it explicitly takes a step. This can be viewed as a limiting-case abstraction corresponding to low-speed, or "friction-dominated," motion. A 300 kg robot moving down a corridor at 3 m/sec no longer satisfies this abstraction. We need an approach to learning control strategies for high-speed, or "momentum-dominated," motion.

High-speed navigation creates demands on the representation of large-scale space distinct from those of low-speed navigation. At high speeds, a robot cannot stop quickly, so in order to avoid collision it must anticipate the environment geometry it will encounter. In general, the greater the range of the environment that the robot anticipates, the faster the robot will be able to safely travel. Since a mobile robot can only sense the part of the environment that is within its sensor range, to travel as rapidly as possible it must use stored knowledge of the environment.

Previous approaches to high-speed navigation can be categorized by the level of spatial knowledge they employ. At one end of this spectrum are methods that assume completely accurate prior knowledge of large-scale space. Using this knowledge and an accurate model of robot dynamics, an approximately time-optimal feedforward control between two locations can be found using a variety of techniques [Gilbert and Johnson 1985, Shiller and Chen 1990, Zaharakis and Guez 1990, Donald and

Xavier 1989]. However, these methods demand a level of model accuracy that cannot be attained in unstructured settings. At the other end of this spectrum, methods described by Feng and Krogh [Feng and Krogh 1990] and Borenstein and Koren [Borenstein and Koren 1989, Borenstein and Koren 1991] do not require a stored map of the environment at all. Instead, the robot navigates to a goal position (expressed in global coordinates) using only currently sensed range information to avoid collision. This eliminates the need for an accurate global map, but since the robot is unable to anticipate environment geometry beyond its sensor range, its performance is restricted. Moreover, the robot may still need to accurately estimate its position in a global frame to avoid critical errors.

We are interested in developing an approach that falls in the middle of this spectrum: though the robot's knowledge about environment geometry is subject to error, the robot can nonetheless use *approximate* knowledge of geometry to enhance its ability to travel at high speed. We are exploring a method for representing knowledge of geometry based upon the spatial semantic hierarchy for this task. We propose an approach to autonomous high-speed navigation that proceeds in the following three stages:

1. **Low-speed exploration.** The environment is explored and mapped at low speed using the exploration strategy described in Section 2.

2. **Initial high-speed control.** The robot combines relatively accurate local range information from sensors with approximate stored knowledge of large-scale space to travel at a higher velocity.

3. **Practice.** The information in the map is refined with experience to enable the robot to travel faster.

5.1   Low-Speed Exploration

The environment is explored and mapped as described in Section 2 with the addition that the uncertainty level in the estimate of the relative positions of distinctive places is described using the approximate transform representation proposed by Smith and Cheeseman [Smith and Cheeseman 1986]. The result of exploring a simple environment is shown in Figure 5.

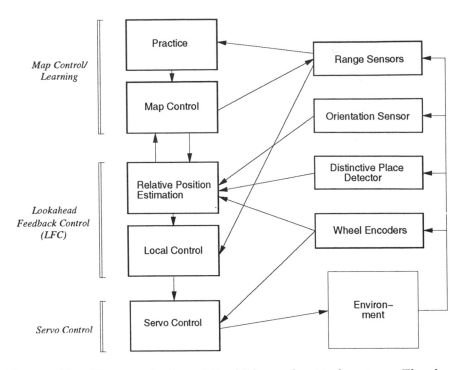

Figure 11. The organization of the high-speed control system. The three brackets indicate groups of components that operate at the same time scale: the servo-control level operates at the fastest time scale; the map-control/learning level at the slowest. In addition to the robot's range and orientation sensors, we assume a *distinctive place detector* that estimates the position of a nearby distinctive place with respect to the robot's frame of reference, and *wheel encoders* that provide an estimate of velocity and a noisy estimate of displacement.

5.2   Initial High-Speed Control

The organization of the high-speed navigation system and the sensors used by the robot is described in Figure 11. The major components can be grouped into three levels, each operating on a successively shorter time scale: the map-control level, the lookahead-feedback-control level (LFC), and the servo-control level.

The Map-Control Level

The map-control level provides the lower levels of the system with information about the large-scale structure of the environment not available from the robot's immediate sensor data. Information needed to rapidly travel a *route* (a sequence of distinctive places and edges) is specified with a sequence of trajectory targets – the approximate position and velocity that the robot should aim for. In experiments to date, all targets are initially located at distinctive places and have zero velocity. To travel toward a distinctive place, the map controller indicates that the *current target* is the target at that distinctive place. The robot travels toward this target without further intervention from the map-control level until the robot gets within some threshold distance of the current target, when the target at the next distinctive place along the route becomes the current target.

The Lookahead-Feedback-Control Level (LFC)

This level must choose a trajectory for each 0.5 - 1.0 second control interval that both satisfies the robot's dynamic constraints and takes the robot to to the current target in minimum time. The first major component of the LFC, the relative-position estimator, uses wheel-encoder and distinctive-place data to maintain an estimate of the robot's position relative to the current target as shown in Figure 12. The local-controller component uses this estimate and range data to choose a trajectory for the next control interval.

The local controller must be able to guide the robot rapidly despite uncertainty about the outcomes of its actions: target position information is approximate, only nearby geometry can be sensed, and the anticipated trajectory may differ from reality due to modeling errors and disturbances. To satisfy these constraints, the local controller employs a depth-limited heuristic search that "looks ahead" a fixed time into the future to find the trajectory for the next control interval that leads to the state on the lookahead horizon with the best current evaluation. State evaluations are the sum of an estimate of the open-space travel time to the current target and a penalty for proximity to obstacles. Uncertainty in the position of the target and the outcome of the robot's control can be handled very easily in this context by basing the travel time estimate on either the expected time to the target, or a worst-case time to the target for a specified confidence level.

The Servo-Control Level

This level compensates for disturbances on a short time scale. Wheel-encoder data is input to a servo controller that attempts to track the trajectory specified by the lookahead-feedback-control level.

5.3   Practice

The robot should be able to use the information it acquires during the initial high-speed traversal of a route to refine its map and travel the route more rapidly. Two straightforward modifications of the map information are possible. First, each traversal between a pair of trajectory targets yields an independent measurement of their relative position which can be used to reduce the uncertainty in this information. Second, the robot can use knowledge of its dynamics to modify the position and velocity of the trajectory targets to be nearer to a point on an optimal trajectory by applying an adaptation of a technique described by Froom [Froom 1991]. The results of applying this method are shown in Figure 13.

5.4   Conclusions

In experiments to date, the implicit representation of approximate map knowledge via trajectory targets has worked surprisingly well, but this is not yet a general method for high-speed navigation. In particular, a method must be developed to introduce trajectory targets *between* distinctive places when collision avoidance at high speed could not otherwise be guaranteed.

6   CONCLUSIONS

This paper reports the current state of our work in applying the semantic hierarchy approach to robot learning about space and action. The

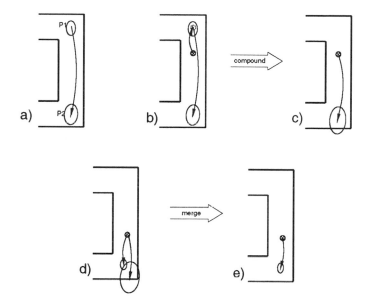

Figure 12. The relative-position estimator uses *compounding* and *merging* operations [Smith and Cheeseman 1986] to maintain an estimate of the relative position of the robot's current target. In (a), the robot is at Place 1, and its approximate knowledge of the relative position of its current target at Place 2 is represented as an approximate transform (AT), here displayed graphically as an arrow (indicating the nominal value of the estimate) and an ellipse (a constant probability contour of the error distribution of the estimate). In (b), after moving part way down the corridor, an estimate of the robots' position relative to Place 1 is represented by another AT, and in (c) these two AT's are *compounded* to obtain a (lower accuracy) estimate of the new relative position of Place 2. As the robot continues down the corridor, this operation is repeated and the uncertainty in the robot's position estimate increases. But when the robot gets close enough to Place 2, as in (d), its distinctive place detector provides an independent estimate of the position of Place 2. These two estimates are *merged* to get a higher accuracy estimate.

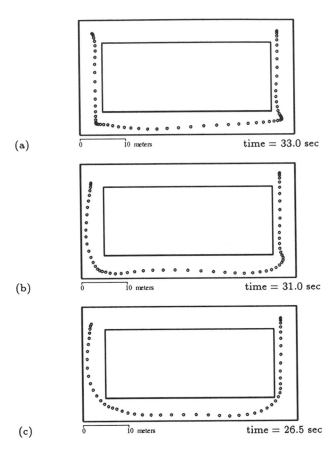

(a)   0   10 meters   time = 33.0 sec

(b)   0   10 meters   time = 31.0 sec

(c)   0   10 meters   time = 26.5 sec

Figure 13. The robot uses the map of the environment created with low-speed exploration to travel along a route using the high-speed control method. The targets initially have zero velocity and are positioned at distinctive places. In (a), the robot stops at each distinctive place. In (b), the robot travels along the route without stopping. After two traversals of the route, estimates of the relative positions of the targets have been improved, and the positions and velocities of the intermediate targets are modified, resulting in the trajectory shown in (c). The robot's sensor range here is 10 meters, its maximum acceleration is $1m/s^2$, and the robot's position is shown at 0.5 second intervals.

approach appears to be a rich and productive one, with far-reaching significance.

From the robotics perspective, our goal is an autonomous robot that can orient itself to its own sensorimotor system through experimentation, learn a map of its environment through exploration, and optimize its high-speed performance through practice. This capability should generalize beyond the immediate application, since spatial knowledge and cognitive maps are frequently used metaphors for other kinds of knowledge and skilled performance (e.g., manipulator control in configuration space).

From the artificial intelligence perspective, the spatial semantic hierarchy demonstrates how a discrete symbolic representation can be used to capture many useful aspects of an agent's continuous interaction with a continuous environment. In this sense, our work supports the Physical Symbol System Hypothesis.

From the cognitive science perspective, the semantic hierarchy approach allows us to build a computational model that expresses the complexity and modular structure of a nontrivial domain of human knowledge. In particular, it supports the position that a complex body of knowledge is acquired, not by a single representation and learning algorithm, but by a highly structured mechanism consisting of several distinct and interacting representations and learning algorithms.

Acknowledgements

Thanks to Boaz J. Super for his helpful comments and discussions, and to the faculty, staff, and students of the Electrical and Computer Engineering Department, especially Dr. Elmer Hixson, Rohan Kulkarni, Jeffrey Meek, Reginald Richard, Kevin Taylor, Richard Vaughan and Lee Yang for their help with the physical robots.

This work has taken place in the Qualitative Reasoning Group at the Artificial Intelligence Laboratory, The University of Texas at Austin. Research of the Qualitative Reasoning Group is supported in part by NSF grants IRI-8905494, IRI-8904454, and IRI-9017047, by NASA grant

NAG 9-512, and by the Texas Advanced Research Program under grant no. 003658-175.

7

UNCERTAINTY IN GRAPH-BASED MAP LEARNING

Thomas Dean
Kenneth Basye
Leslie Kaelbling

*Department of Computer Science,
Brown University, Providence, RI 02912*

ABSTRACT

For certain applications it is useful for a robot to predict the consequences of its actions. As a particular example, consider programming a robot to learn the spatial layout of its environment for navigation purposes. For this problem it is useful to represent the interaction of the robot with its environment as a deterministic finite automaton. In map learning the states correspond to *locally distinctive places*, the inputs to robot actions (navigation procedures), and the outputs to the information available through observation at a given place. In general, it is not possible to infer the exact structure of the underlying automaton (*e.g.*, the robot's sensors may not allow it to discriminate among distinct structures). However, even learning just the *discernible* structure of its environment is not an easy problem when various types of uncertainty are considered. In this chapter we will examine the effects of only having probablistic information about transitions between states and only probablistic knowledge of the identity of the current state. Using this theoretical framework we can then determine whether it is at all possible for a given robot to learn some specific environment and, if so, how long this can be expected to take.

1 INTRODUCTION

We assume that for certain applications it is useful for a robot to predict the consequences of its actions. The actions under consideration might be navigational, observational, or manipulatory. They might affect the position and orientation of the robot or other nearby objects, or the

171

robot's knowledge of such positions and orientations. In this chapter, we are concerned with learning to predict the *state* of the robot following a sequence of discrete actions.

In general, we are interested in having a robot infer a model of the dynamics of its interaction with its environment. However, as motivation, we will frequently refer to the problem of programming a robot to learn the spatial layout of its environment for navigation purposes. As it may not be immediately apparent that spatial reasoning and navigation can be described in terms of action sequences and discrete state changes, we begin with some background on *qualitative navigation*.

2   QUALITATIVE NAVIGATION AND MAP LEARNING

Given adequate perceptual capabilities, it is possible for a robot to partition the space which comprises its physical environment into a relatively small set of equivalence classes, so that it can determine the class of the location which it occupies without a great deal of effort. Conceptually, these equivalence classes coarsely tessellate space into regions that are homogeneous by some perceptual criterion. Moreover, it is often possible to design robust procedures that enable a robot to efficiently navigate through such perceptually homogeneous regions.

As a concrete example, consider a mobile robot given the task of moving about in an office-like environment. In such an environment, corridors are easily distinguishable from the junctions in which two or more corridors meet using simple acoustic or infrared ranging sensors; ignoring orientation, all junctions where exactly two corridors meet at a right angle are perceptually equivalent to one another. Kuipers [Kuipers 1978] refers to spatial features such as corridor junctions in office environments or intersections in city environments as *locally distinctive places*(LDPs).[1] Ignoring length, straight corridors are also perceptually equivalent, but we generally treat them as distinct from LDPs. Corridors primarily serve

---

[1][Levitt *et al.* 1987] demonstrate how the notion of locally distinctive place can be extended to deal with large open areas.

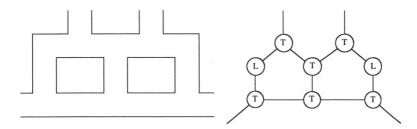

Figure 1. Spatial modeling with labeled graphs

to enable a robot to move from one LDP to another, and we refer to them as *conduits* to emphasize this role.

We assume that the robot has the capability to identify the equivalence class of an LDP (*e.g.,* is it an L- or T-shaped junction) and to traverse conduits. The LDPs and conduits in a given environment induce a labeled graph in which the vertices correspond to LDPs, the edges to conduits, the label on an LDP indicates its equivalence class, and the label on a conduit indicates a particular navigation procedure for traversing the conduit. Figure 1 provides a simple example of a labeled graph induced from an office environment. Learning a spatial environment, which we call *map learning*, corresponds to inferring the labeled graph.

Map learning is performed to expedite point-to-point navigation and assist in path planning. The basic method of inducing a graph from the perceptual and navigational capabilities of the robot results in compact spatial representations, efficient learning, and a simple modular approach to robot software design. While such graph models are used primarily to make qualitative spatial distinctions, metric annotations can easily be added to provide more information to guide path planning.

In the process of map learning, the robot actively explores its environment, making observations, and in some cases determining where to look next on the basis of what it has observed so far. We assume that the navigation and observations procedures are given. The problem of learning what constitutes a locally distinctive place is quite interesting but beyond the scope of this chapter.

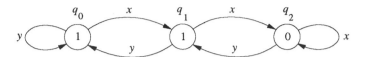

Figure 2. State-transition graph for a finite automaton

It should be noted that map learning is just a special case of learning a model for a finite-state, discrete-time dynamical system. In the following section, we introduce a general mathematical framework and cast learning such models in terms of learning finite automata. This general framework will allow us to draw upon a wide range of theoretical results that pertain to the problem of learning models for dynamical systems.

3   THEORETICAL DEVELOPMENT

We represent the interaction of the robot with its environment as a deterministic finite automaton (DFA). For the map learning problem, the states correspond to LDPs, the inputs to robot actions (navigation procedures), and the outputs to the information available through observation at a given LDP. A DFA is a six tuple, $M = (Q, B, Y, \zeta, q_0, \gamma)$, where

- Q is a finite nonempty set of states,

- B is a finite nonempty set of inputs or basic actions,

- Y is a finite nonempty set of outputs or percepts,

- ζ is the transition function, $\zeta : Q \times B \to Q$,

- q_0 is the initial state, and

- γ is the output function, $\gamma : Q \to Y$.

The *state-transition* graph for a simple deterministic finite automaton is shown in Figure 2. Let $A = B^*$ denote the set of all finite sequences of actions, and $|a|$ denote the length of the sequence $a \in A$. Let $q\langle a \rangle$ be the sequence of outputs of length $|a| + 1$ resulting from executing the sequence a starting in q, and qa be the final state following the execution of the sequence a starting in q. For the automaton shown in

Figure 2, $q_0\langle x^3\rangle = 1100$. The task of learning is to infer the structure of the underlying automaton of the environment by performing a sequence of basic actions and observing the outputs at each step.

In general, it is not possible to infer the exact structure of the underlying automaton (*e.g.*, the robot's sensors may not allow it to discriminate among distinct structures). As a result, we will be satisfied if the robot learns the *discernible* structure of its environment. An automaton is said to be *reduced* if, for all $q_1 \neq q_2 \in Q$, there exists $a \in A$ such that $q_1\langle a\rangle \neq q_2\langle a\rangle$. A reduced automaton is used to represent the *discernible* structure of the environment; you cannot expect a robot to discern the difference between two states if no sequence of actions and observations serves to distinguish them.

In this chapter, we are concerned with the effect of uncertainty on learning. Uncertainty can arise due to there being incomplete information available through observations at a state, or through errors in perception regarding the output at a state or the actual movements resulting from executing navigation procedures.

In the remainder of this chapter, we distinguish between the output function, γ, and the *observation* function, φ. We say that the output function is *unique* if $\forall q_1, q_2 \in Q, \gamma(q_1) = \gamma(q_2)$ implies $q_1 = q_2$; otherwise, it is said to be *ambiguous*. If the output function is unique and $\varphi = \gamma$, then there is no uncertainty in observation and learning is easy. If the output function is ambiguous, then even if $\varphi - \gamma$, learning is hard without further restrictions on the problem. We will be more precise regarding just how 'hard' in Section 5.

To facilitate learning, it is often useful for the robot to determine where it is (*i.e.*, determine its current state) or find its way to some state that it can distinguish from all others. A *reset* allows the robot to return to the initial state at any time. We generally assume that the underlying state-transition graph is connected; this is particularly important if we require the robot to learn the entire state-transition graph without a reset. The availability of a reset provides a powerful advantage by allowing the robot to anchor all of its observations with respect to a uniquely distinguishable state, q_0.

For a finite automaton, the ability to move to some state that is distinguishable from all other states is almost as powerful as a reset. A *homing sequence* provides the robot with the ability to do just that. A homing sequence, h, allows the robot to distinguish the states that it ends up in immediately following the execution of h; the sequence of observations $q\langle h \rangle$ constitutes a unique signature for state qh. A homing sequence, $h \in A$, has the property that, for all $q_1, q_2 \in Q$, $q_1\langle h \rangle = q_2\langle h \rangle$ implies $q_1 h = q_2 h$. Every reduced automaton has a homing sequence; however, the shortest homing sequence may be as long as $|Q|^2$ [Rivest and Schapire 1989].

There is also something that is almost as powerful as having an output function that is unique. A sequence, $d \in A$, is said to be a *distinguishing sequence* if, for all $q_1, q_2 \in Q$, $q_1\langle d \rangle = q_2\langle d \rangle$ implies $q_1 = q_2$. (Every distinguishing sequence is a homing sequence, but not the other way around.) A distinguishing sequence, d, allows the robot to distinguish the states that it starts executing d in; the sequence of observations $q\langle d \rangle$ constitutes a unique signature for q. The single action x is a distinguishing sequence for the automaton shown in Figure 2. Given a distinguishing sequence for the underlying automaton, if $\varphi = \gamma$, then learning is easy. Not all automata have distinguishing sequences.

For the case in which $\varphi \neq \gamma$, we assume there is a probability distribution governing what the robot observes in a state. We will confine our attention to the case in which each visit to a state is an independent trial.[2] To avoid pathological situations, we assume that the robot observes the actual output with probability better than chance; that is,

$$\forall q \in Q, \Pr(\varphi(q) = \gamma(q)) \geq \alpha > 0.5,$$

where, in this case, $\varphi(q)$ is a random variable ranging over Y. The closer α is to $1/2$, the less reliable the robot's observations.

To model uncertainty in movement, we introduce a *navigation* function, η, distinct from but related to the state-transition function, ζ. If $\eta = \zeta$, then there is no uncertainty in movement. As in the case of observation,

---

[2] We ignore the case in which the robot remains in the same state by repeating the empty sequence of actions and observes the output sufficiently often to get a good idea of the correct output. In this case, the independence assumption regarding different observations of the same state is not even approximately satisfied.

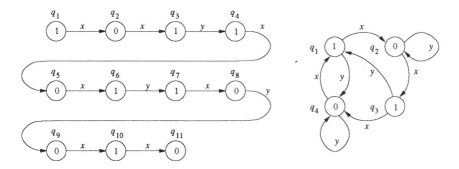

Figure 3. Two automata consistent with the input

to avoid pathological situations, we assume that the robot's actions take it to the states determined by the state transition function with better than chance; that is,

$$\forall a \in B, q \in Q, \Pr(\eta(q,a) = \zeta(q,a)) \geq \beta > 0.5,$$

where, in this case, $\eta(q,a)$ is a random variable ranging over Q.

Given our assumptions regarding η and φ, the robot's interaction with its environment is largely determined by the underlying DFA defined by the functions ζ and γ. We are primarily concerned with learning this underlying deterministic automaton and will generally ignore the problems involved in learning the probabilistic transitions and observations.

There are different sorts of feedback that a robot might receive in the process of learning. The robot might accept input from a teacher, it might learn while performing some other task, or it might actively explore its environment using a random or directed strategy. A teacher might tell the robot where to explore. A particularly useful teacher might tell the robot if it gets the automaton right or, if it is wrong, point out exactly how it is wrong by providing a counterexample where the robot's automaton predicts the wrong outcome.

We are also interested in learning automata efficiently. By efficiently, we mean that the time spent in exploration is at most polynomial in the size of the problem and some measure of how difficult the problem is. Difficulty might be measured in terms of how close α and β are to $\frac{1}{2}$, or, in the case of problems that provide probabilistic guarantees, in terms of how close the guaranteed probability of success is to one. If we can prove such a polynomial bound then we say that the problem is *polynomial-time learnable*. In practice, we want low-order polynomial or even linear-time algorithms. We also want the robot to learn a compact representation of its environment. Figure 3 shows an input sequence on the left; both the input and the structure on the right correspond to automata that are consistent with the input. For the most, part we will be interested in inferring the smallest automaton consistent with the observed input.

4   PROBLEM CLASSIFICATION

There are a number of features that distinguish the problems that we are interested in this chapter. We have already mentioned some of those features: unique versus ambiguous output functions, perfect ($\varphi = \gamma$) versus imperfect ($\varphi \neq \gamma$) observations, perfect ($\eta = \zeta$) versus imperfect ($\eta \neq \zeta$) transitions, the existence of resets and distinguishing sequences. While the output function may, in general, be ambiguous, there are often cases in which the output available in a given state is distinguishable from all other states; we call such states *landmarks* alluding to the spatial case. Environments liberally sprinkled with landmarks are often easier to learn than those environments without landmarks.

Other features of interest concern the robot's ability to move about reliably. If the robot can always reverse its last action to return to its previous state, then learning can be simplified in some cases even if the robot is not aware of which action is required to reverse its steps. Such *reversible* environments allow us to make use of certain desirable computational properties of undirected graphs.

In the following section, we provide a number of theoretical results. Ideally, we would like to provide an efficient algorithm that solves the most general version of the learning problem. Unfortunately, the general problem is provably hard, and so we play a complicated game specializing the general problem to obtain positive results. If we can't find an efficient solution for a given problem, then we try to prove that it is hard. If we

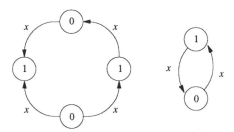

Figure 4. Two perceptually indistinguishable automata

succeed in proving that the problem is hard, then we try to find interesting restricted versions of the problem that are easy. The restrictions might involve a more transparent environment, a more powerful teacher, or a less powerful adversary in the case of adversary arguments. If we succeed in finding an efficient solution for a problem, then we might try to lessen the restrictions somewhat to solve a more general problem. The object is to understand just what features of these problems in isolation or combination make the problem easy or hard.

5   SUMMARY OF RESULTS

This section summarizes a variety of theoretical results. We describe problems, comment on associated computational issues, and in some cases provide a related theorem along with a reference for where a proof can be found in the literature.

5.1   Deterministic Transitions and Observations

In general, it is not possible for a robot to recover the complete structure of its environment (*e.g.,* Figure 4 shows two automata that are perceptually indistinguishable). For this reason, we primarily restrict our attention to learning reduced automata. Provided with some additional capabilities, however, it is possible to learn the complete structure. Dudek *et al.* [Dudek *et al.* 1991] provide a polynomial-time algorithm for recovering the complete structure of an environment given one or more markers that the robot can put down and pick up at states which it visits in its explorations. In the following, we do not allow the use of markers and assume that the automaton is reduced.

The case in which the output function is ambiguous and $\varphi = \gamma$ has been studied extensively. The results in the literature are often stated in terms of learning an automaton that accepts particular language. Let $L(A)$ be the language accepted by the automaton A. We can partition the set of states into accepting ($\gamma(q) = 1$) and nonaccepting ($\gamma(q) = 0$) states, so that $\gamma(q_0 x) = 1$ iff $x \in L(A)$. In exploring its environment, the robot can be viewed as making *membership* queries corresponding to asking whether or not a given string x is in $L(A)$ where A is the underlying automaton. The answers returned from these queries can be divided into positive ($x \in L(A)$) and negative ($x \notin L(A)$) examples. At any point during its exploration, the robot can be seen as trying to determine the smallest DFA consistent with the examples seen thus far. This problem turns out to be difficult.

Theorem 1 *The problem of finding the smallest DFA consistent with a set of positive and negative examples is NP-hard [Gold 1972].*

Even finding a DFA polynomially close to the smallest is intractable assuming P$\neq$NP.

Theorem 2 *For any fixed polynomial p and set of positive and negative examples S, the following problem is NP-hard: find a DFA consistent with S and of a number of states at most $p(n^*)$, where n^* is the minimum number of states of any DFA consistent with S [Pitt and Warmuth 1989].*

Perhaps, you might ask, it is possible to make just the right membership queries and learn an automaton efficiently. Unfortunately, this is not the case.

Theorem 3 *DFAs are not polynomial-time learnable using membership queries only [Angluin 1978].*

There are, of course, more powerful sorts of queries that will allow us to learn automata efficiently. For instance, suppose that the robot is allowed to ask *state-transition* queries in which the robot provides a string x and receives in return $q_0 x$, the state that results from executing the string in the initial state. Learning with state-transition queries turns out to be easy.

Theorem 4 *DFAs are polynomial-time learnable using state-transition queries only [Tzeng 1992].*

It should be noted, however, that the corresponding problem of finding the smallest DFA consistent with a set of examples of the form $\langle x, q_0 x \rangle$ is NP-complete [Tzeng 1992]. State-transition queries are not all that interesting for robot learning, since the robot generally cannot distinguish states perceptually. Another type of query that has been studied is called the *equivalence* query. In an equivalence query, the robot hypothesizes that the automaton is A' and asks if $L(A) = L(A')$; if not, the robot receives in return a string x, called a *counterexample*, such that x is in one of $L(A)$ or $L(A')$ but not the other. Equivalence queries alone are not sufficient for polynomial-time performance.

Theorem 5 *DFAs are not polynomial-time learnable using equivalence queries only [Angluin 1978].*

However, Angluin, building on the work of Gold [Gold 1972], provides a polynomial-time algorithm for inferring the smallest DFA given the ability to make both membership and equivalence queries.

Theorem 6 *DFAs are polynomial-time learnable using membership and equivalence queries [Angluin 1987].*

Angluin's algorithm depends on the ability to reset the automaton to the initial state. The notion of a reset is unnatural for many applications including those involving map learning. Without a reset and with no way of determining when the robot is in the initial state, the language framework that we introduced earlier is no longer an appropriate model. From now on, we focus on the model of the robot exploring its

environment by traversing the directed graph associated with the underlying DFA. Membership queries are replaced with observation sequences
resulting from executing specific action sequences and the counterexamples from equivalence queries are replaced with observations that differ
from those predicted by the robot's hypothesized automaton. Rivest
and Schapire show how to dispense with the reset in the general case
[Rivest and Schapire 1989], and how to dispense with both the reset
and the source of counterexamples in the case in which a distinguishing
sequence is either provided or can be learned in polynomial time [Rivest
and Schapire 1987, Schapire 1991].

This last result is particularly important for the task of learning maps.
For many man-made and natural environments it is relatively easy to
find a distinguishing sequence. In most office environments, a short,
randomly chosen sequence of turns will serve to distinguish all junctions
in the environment. Large, complicated mazes do not have this property,
but we are not interested in learning such environments. It is easy to
prove that distinguishing sequences make learning DFAs easy.

Theorem 7 *DFAs are polynomial-time learnable using directed exploration if the robot is given a distinguishing sequence of length polynomial
in the size of the DFA.*

If γ is unique, then the null action sequence is a distinguishing sequence,
and there is a one-to-one correspondence between Y and Q; this special
case of Theorem 7 follows directly from Theorem 4. It would be nice
if, knowing that an environment has a distinguishing sequence, a robot
could proceed to find it. Unfortunately, the problem is still hard if we
restrict our attention to DFAs with distinguishing sequences.

Theorem 8 *There is no polynomial-time algorithm to learn the class
of DFA environments with distinguishing sequences by exploration only,
even if the goal is identification with high probability and a reset operation
is available [Dean et al. 1992].*

In all of the above problems, movement is certain and $\varphi = \gamma$. The only
uncertainty arises in cases in which the output function is not unique.

If we allow for probabilistic errors in state transition and observation, then we are interested in inferring *stochastic automata* (probabilistic transitions only) and *Markov chains* (probabilistic transitions and observations). The latter appear in the literature under the heading of *hidden Markov models* [Levinson *et al.* 1983]. The work on hidden Markov models generally assumes that the underlying automata can be reset to its initial state and is primarily concerned with learning in the limit. Rudrich [Rudrich 1985] provides algorithms for inferring the graph structure and transition probabilities of Markov chains in the limit. Tzeng [Tzeng 1992] considers the problem of exactly learning stochastic automata in polynomial time assuming a learning model similar to Angluin's [Angluin 1987] and relying on a variety of different query types including membership, equivalence, and state-transition queries.

Probabilistic transitions and observations complicate the problem in that the robot can never be certain about what it sees or what actual transitions it made in carrying out an action. As a result, we consider algorithms with probabilistic guarantees. Typically, we wish to guarantee that the robot exactly learns the underlying DFA with probability $1 - \delta$ in time polynomial in the size of the DFA, $\frac{1}{\delta}$, and perhaps some other factors.

5.2 Deterministic Transitions and Stochastic Observations

It would simplify the problem if the robot was able to determine its location with arbitrarily high probability. Dean *et al.* [Dean *et al.* 1992] provide an algorithm for the case with probabilistic observations but deterministic transitions which exploits the determinism in movement to determine the robot's location with high probability.

Theorem 9 *It is possible to learn any DFA A with probability $1 - \delta$ given a distinguishing sequence for A in time polynomial in $\frac{1}{\alpha - \frac{1}{2}}$, $\frac{1}{\delta}$ and the size of A, where α is a lower bound over all states on the probability of successfully observing the output at a state [Dean et al. 1992].*

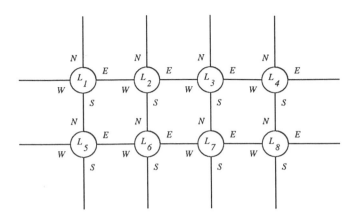

Figure 5. A tessellation of order four

The more general case of probabilistic transitions and observations without more powerful queries is largely open at this point. In the following, we describe a number of special cases. The problems arising from probabilistic transitions appear to cause the most significant problems. All of the following problems involve probabilistic transitions.

5.3 Stochastic Transitions and Deterministic Observations

The class of DFAs in which there is no uncertainty in observation ($\varphi = \gamma$ and γ is unique) and the underlying graph is a regular tessellation (every vertex has the same degree and there are no self transitions) and effectively undirected (if there is an edge from q to q' then there is an edge from q' to q) is polynomial-time learnable. Figure 5 shows a regular tessellation of order four; at each location, the robot can move in one of four directions. In this case, the robot's information about how it has moved is very limited. Once it moves, it has no way of knowing which direction it went, and no way of knowing which direction it has come from when it arrives at its new location.

Theorem 10 *It is possible to completely learn any finite regular tessellation $G = (V, E)$ with probability $1 - \delta$ in time polynomial in $\frac{1}{\delta}$ and the size of G [Basye et al. 1989].*

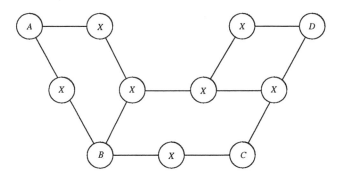

Figure 6. A landmark graph with distribution parameter of two

The next problem we look at involves probabilistic transitions and deterministic observations with an output function that is unique only over some subset of states: the set of landmarks. We speak in terms of learning the state-transition graph for the underlying automaton. Think of each edge in the graph as being labeled with the *direction* to take in order to move to the vertex incident with the other end of the edge. We consider the case in which movement in the intended direction takes place with probability better than chance, and that, upon entering a vertex, the robot knows with certainty the local name of the edge upon which it entered. We call the latter requirement *reverse movement certainty*. This requirement provides one way of bounding the number of false positives (the robot thinking it has successfully negotiated a path when it has not) and false negatives (the robot thinking it has failed to negotiate a path when it has), Without any means of bounding such errors, learning becomes impossible.

At any point in time, the robot is facing In a direction defined by the label of a particular edge/vertex pair—the vertex being the location of the robot and the edge being one of the edges emanating from that vertex. We assume that the robot can turn to face in the direction of any of the edges emanating from the robot's location. We also assume that upon entering a vertex the robot can determine with certainty the direction in which it entered. Directional uncertainty arises when the robot attempts to move in the direction it is pointing. Let $\beta > 0.5$ be the probability that the robot moves in the direction it is currently pointing. More than 50% of the time, the robot ends up at the other end of the edge defining its current direction, but some percentage of the time it ends up at the other end of some other edge emanating from its starting vertex. While the robot won't know that it has ended up at some unintended location, it will know the direction to follow in trying to return to its previous location.

With regard to observations, we assume that the locations in the world are of two kinds, those that can be distinguished, and all others. That is, there is some set of landmarks, in the sense explained above, and all other locations are indistinguishable. We model this situation using a partitioning W of V and assuming that we have a sensor function which maps V to W. Here W consists of some number of singleton sets plus the set of all indistinguishable elements. We further assume that a second sensor function allows us to determine whether the current location is or is not a landmark. For convenience, we define D to be the subset of V consisting of the landmark vertices and I to be the subset of V consisting of the non-landmark vertices. We refer to this kind of graph as a *landmark graph*. We define the *landmark distribution parameter*, r, to be the maximum distance from any vertex in I to its nearest landmark (if $r = 0$, then I is empty and all vertices are landmarks). We say that a procedure learns the *local connectivity within radius r* of some $v \in D$ if it can provide the shortest path between v and any other vertex in D within a radius r of v. Figure 6 depicts a graph with a landmark distribution parameter of two. We say that a procedure learns the *global connectivity of a graph G within a constant factor* if, for any two vertices u and v in D, it can provide a path between u and v whose length is within a constant factor of the length of the shortest path between u and v in G.

Thus, we may summarize the robot's capabilities as follows. The robot's movement function is not perfect, but serves to move the robot in the intended direction more than half the time. At each vertex, the robot knows how many out-edges there are, and what their labels are. In addition, the robot knows which of these it has just arrived from, whether the vertex is a landmark, and if so, what its unique name is. These requirements — particularly the one involving reverse certainty — are quite restrictive, but they enable us to obtain the following theorem.

Theorem 11 *It is possible to learn the global connectivity of any landmark graph with probability* $1 - \delta$ *in time polynomial in* $\frac{1}{\beta - \frac{1}{2}}$, $\frac{1}{\delta}$, *and the size of G, and exponential in r [Basye et al. 1989].*

Note that in the case in which the output function is unique (landmark parameter of one) the exponential factor disappears. Even discounting the exponential factor, the last result is rather restrictive. Cases in which transitions are probabilistic appear to be resistant to polynomial solutions. In light of this, it might be a good idea to change our game. One way to change the game is to make what we will call *distributional assumptions*. In the previous problem, we made assumptions regarding the density of landmarks; this is an example of a distributional assumption. For obvious reasons, we prefer distributional assumptions that have some physical motivation.

To illustrate the utility of distributional assumptions, we introduce a very simple algorithm for learning graphs, and provide distributional assumptions for which the algorithm can be shown to satisfy a particular performance guarantee.

5.4   Stochastic Transitions and Observations

The basic problem we are concerned with involves an environment with probabilistic transitions and observations but with a unique output function. Given the unique output function, we can assume that each state, q_i, has an associated, unique label, l_i. Let P_{ij} be the probability of observing label l_i given that the robot is in the state q_j. In a given state, the robot will have observed some label l_i, and, following the execution of some action x, it will observe some label l_j; we will be concerned with obtaining a sufficient number of such pairs of labels. The robot employs a random strategy for exploring its environment; that is, it chooses from among the actions that it has in a given state (location) according to a uniform distribution. As in the case of the regular graphs, we assume that the graph is effectively undirected.

The algorithm that we consider here uses a random strategy to explore the graph and records, for each pair of labels, (l_i, l_k), and each action, x,

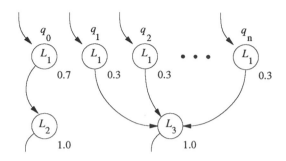

Figure 7.   Fragment of a confusing automaton

the number of times that an observation of l_i followed by performing x
resulted in an observation of l_k. After enough exploration, a graph can
be inferred from these statistics as follows. If l_k is the most frequently
observed label after doing action x when observing label l_i, then we
assume that there is an edge for action x from state q_i to state q_k in
the underlying graph. Now we consider cases in which this algorithm is
likely to work.

Map learning is difficult if there are some locations that are easily confused with several other locations. Suppose that q_0 is observed to have
label L_1 70% of the time, each of $q_1, q_2, \ldots, q_n$ are observed to have label
L_1 30% of the time, $q_0 x$ is observed to have label L_2 100% of the time,
and each of $q_i x$ for $1 \le i \le n$ are observed to have label L_3 100% of the
time. Figure 7 shows a fragment of state-transition graph illustrating the
situation. It is likely for $n > 2$, that the robot will observe L_1 followed
by L_3 more frequently than L_1 followed by L_2, and, hence, incorrectly
guess the structure of the underlying automaton. We can avoid such
situations by bounding the probability that the robot is mistaken about
its being in a location. To ensure this requirement, it is sufficient that
$\forall i, P_{ii} > \sum_{j \ne i} P_{ij}$. The requirement is also satisfied if $\forall i \ne j, P_{ij} = P_{ji}$.

Map learning is also made more difficult if the robot has no idea of what
distribution it is sampling from. This is particularly apparent when we
consider the frequency with which the robot visits different locations.
Suppose that q_1 is observed to have label L_1 70% of the time, q_2 is
observed to have label L_1 30% of the time, $q_1 x$ is observed to have label
L_2 100% of the time, and $q_2 x$ is observed to have label L_3 100% of

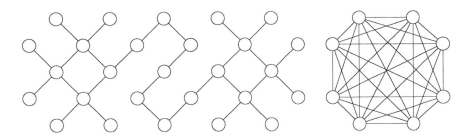

Figure 8. Graphs with low and high conductance

the time. Now, suppose that the robot visits q_2 three times as often as it visits q_1; again, the robot will incorrectly guess the structure of the underlying automaton. If we knew the distribution governing the frequency of visits to locations, then we could avoid these problems. In the case of deterministic transitions and probabilistic observations described above, the robot exploited the determinism in movement to establish with high probability a stable basis for sampling. In the case of probabilistic transitions and observations, the problem is more difficult.

Ideally, we would like the robot to visit locations according to a uniform distribution. If the robot is aware of the distribution governing its errors in movement, then the robot can explore so that its behavior corresponds to a Markov chain with stationary distribution that is uniform. Think of the robot as issuing commands that result in actions. The actions are deterministic while the results of commands are stochastic. Let $\Pr(a|c)$ be the probability that a command c will result in an action a, and suppose that this probability is independent of the robot's starting state. If the robot knows $\Pr(a|c)$ for all a and c, then it can perform a walk corresponding to a Markov chain whose transition matrix is symmetric and doubly stochastic thus guaranteeing a uniform stationary distribution.

The only problem remaining concerns how long a walk is necessary before the sampled distribution (corresponding to the state visitation frequencies for a walk of a specific length) is reasonably close to the stationary distribution. We are guaranteed that the stationary distribution is uniform in the limit, but, as yet, we have no idea how long it will take to reasonably approximate this distribution. It turns out that how long it will take will depend on the *conductance* of the edge-weighted state-transition graph in which the weight of an edge corresponds to the probability of traversing that edge in a random walk. Informally, the conductance, Φ, of a graph is a measure of how many ways there are to get from one part of the graph to another. In Figure 8, the graph on the left has relatively low conductance while the graph on the right has relatively high conductance, assuming in each case that the edges emanating from a given vertex are assigned an equal weight or transition probability. Mihail [Mihail 1989] provides a convenient characterization of the relationship between the conductance of a graph, the length of a random walk governed by a Markov chain, and the discrepancy between the sampled and stationary distributions. Now we can state (with some simplification to avoid introducing additional mathematics) the following theorem.

Theorem 12 *The output of the algorithm described above is correct with probability at least $1 - \delta$ after a uniform random walk of length polynomial in $\frac{1}{\alpha - \frac{\sqrt{2}}{2}}$, $\frac{1}{\beta - \frac{1}{2}}$, $\frac{1}{\delta}$ $\frac{1}{\Phi}$, and the size of the underlying automaton, if $\forall i \neq j, P_{ij} = P_{ji}$ and the underlying automaton is effectively undirected [Kaelbling et al. 1992].*

As should be apparent, the story becomes more complex as we try to find some reasonable tradeoff between performance and generality. We want to be able to handle noise in observation and movement, but at the same time we have to avoid pathological situations. Our intuition is that there are reasonably 'benign' environments for which map learning is easy. The trick is characterizing such environments mathematically and then devising algorithms with satisfactory performance.

6   CONCLUSIONS

In this chapter, we explore theoretical results in learning the structure, spatial and otherwise, of environments whose important structure can be represented as labeled graphs. Our objective here is foundational; we are trying to understand the fundamental sources of complexity involved in learning about the structures induced by perception. Many researchers believe that map learning should be easy in relatively benign, non maze-like environments; we would like to explain why this is the case from a computational perspective. Practically speaking, the running times provided in the proofs for the theorems described in this chapter are unacceptable; ultimately, we are interested in algorithms whose effort in relatively benign environments is bounded by some small constant factor of the size of the environment.

There is a great deal of work left to be done involving problems in which both movement and observations are prone to stochastic errors. With regard to map learning, it should be noted that we have made no use of the fact that the graphs we are dealing with are planar and very well behaved. It should also be noted that our notion of correctness is rather unforgiving; in most real-world applications, it is seldom necessary to learn about the entire environment as long as the robot can navigate efficiently between particular locations of interest.

With regard to more forgiving measures of performance, it may be possible to extend the techniques described in this chapter to find ϵ-approximations rather than exact solutions by using the *probably approximately correct* learning model of Valiant [Valiant 1984]. Such extensions would require the introduction of a distribution, Pr, governing both performance evaluation and exploration. As one simple example, suppose that the robot operates as follows. Periodically the robot is asked to execute a particular sequence of actions chosen according to Pr; we assume that, prior to being asked to execute the sequence, it has determined its location with high probability and that immediately following it executes the distinguishing sequence. Performance is measured in terms of the robot's ability to plan paths between locations identified by their distinguishing sequence signatures, where the paths are generated according to Pr. The problem with this approach is that the starting locations of the sequences are not determined solely by Pr. In fact, all that Pr governs is $\Pr(x|q_{start})$, where q_{start} is the start state and x is some sequence of actions. In order for the above approach to work, the marginal distribution governing the possible starting locations of the robot, $\Pr(q_{start})$, must be the same for both training and performance evaluation.

The general area of learning dynamics and the specific area of learning maps are of broad practical and theoretical interest. In this chapter, we have concentrated on applying the tools of computational learning theory to develop efficient (in the sense of polynomial time) algorithms. There is also a growing literature that concentrates on empirical studies of the same problems [Dean *et al.* 1990, Engelson and McDermott 1992, Kuipers and Byun 1988, Mataric 1990a]. It is hoped that a combination of empirical and theoretical investigations will yield deeper insight into the computational issues in the near future.

Acknowledgements

This work was supported in part by a National Science Foundation Presidential Young Investigator Award IRI-8957601, by the Air Force and the Advanced Research Projects Agency of the Department of Defense under Contract No. F30602-91-C-0041, and by the National Science foundation in conjunction with the Advanced Research Projects Agency of the Department of Defense under Contract No. IRI-8905436.

8

REAL ROBOTS,
REAL LEARNING PROBLEMS

Rodney A. Brooks
Maja J. Mataric

Artificial Intelligence Laboratory,
Massachusetts Institute of Technology, Cambridge, MA 02139

ABSTRACT

The weaknesses of existing learning techniques, and the variety of knowledge necessary to make a robot perform efficiently in the real world, suggest that many concurrent, complementary, and redundant learning methods are necessary. We propose a division of learning styles into four main types based on the amount of built-in structure and the type of information being learned. Using this classification, we discuss the effectiveness of various learning methodologies when applied in a real robot context.

1 INTRODUCTION

For over thirty years AI researchers have harbored the dream of having their programs and robots learn for themselves. According to the most optimistic scenario, human programmers would by now have accomplished enough to program themselves out of a job [Widrow 1962].

The goal of learning in a robot is to prepare it to deal with with unforeseen situations and circumstances in its environment. The refrain that "it can only do what it was programmed to do" no longer applies, as the robot must literally be programmed to do more than it was programmed to do.

The fact that even the simplest of animals seem to be very adaptable suggests that since learning is universal in the animal world, it must be important for survival. By analogy, learning may be equally important in the construction of robust robots. But in order to determine what types of information should be learned and what should be built-in, we must analyze the benefits and the costs of providing a robot with a learning capability.

In this paper we will review the different types of information a robot needs to learn, classify it based on how what is learned translates into action, and suggest learning methods appropriate for each of the resulting categories. Before we approach the problem of robot learning, however, we will consider the questions of cost and benefit in the case of learning in animals, and try to draw relevant analogies and implications.

2   MOTIVATION

There are two main benefits of learning in biological systems. First, learning lets the animal adapt to different circumstances in the world, giving it a wider range of environmental conditions in which it can operate effectively. Second, learning reduces the amount of genetic material and intermediate structures required for building the complete functioning adult animal. In some circumstances it is simpler to build a small structure capable of constructing a larger one, than to specify the larger structure directly.

The first rationale for learning in animals directly applies to robots as well. We would like our robots to adapt to changing external circumstances (e.g. changes in terrain, lighting, or actions and density of other robots), adapt to changing internal circumstances (e.g. drift in sensors and actuators, unexpected sensor or actuator failure, or partial loss of power), and to perform new tasks when appropriate. But if the robot is to adapt to unforeseen circumstances, it must use representational structures rich enough to accommodate the entire variety of possible knowledge to be learned.

The second rationale for learning in animals does not transfer directly to robot control, unless they are programmed using genetic techniques (see [Brooks 1991a] for a discussion of the issues). For instance, many topographical maps within animal brains are built through learning during development instead of being hard-wired genetically. A topographic map is one that preserves locality from sensor space to the neural level (e.g. touch sensors on adjacent patches of skin, or adjacent light sensors in the retina). It is very unlikely that a genetic program that specifies all such connections in detail would evolve, especially since there is little evolutionary pressure for correcting small errors. Instead, animals develop largely undifferentiated sheaths of nerve bundles connecting sensors to the brain site, then use local adaptation based on external or internally generated stimuli to determine the connections.

Unlike biological systems, the necessity for learning in robots is not obvious. Depending on the particular robot and task, the necessary information can be built-in or computed directly. However, much research effort has been devoted to *tabula rasa* learning techniques that assume no built-in information. Although learning from scratch may constitute an interesting intellectual challenge, it poses an unreasonably difficult problem. Indeed, there is no known precedent in biology for such unstructured learning.

Based on the amount of built-in structure, we can group the existing methodologies into two main categories:

- *Strong Learning.* In this approach, nothing is predefined and everything must be learned. [Drescher 1991] is an example of this approach in simulation and [Kaelbling 1990] gives an example of a physically embodied robot. Although in theory such robots should be able to learn everything, in practice it is difficult for them to learn anything.

- *Weak Learning.* In this approach the robot has many built-in capabilities or much *a priori* knowledge, which need not be explicitly or declaratively represented. The structure of the built-in knowledge restricts what the robot can learn.

So far, strong learning approaches have led only to weak results, while weak learning approaches have been much more successful. We propose a compromise that combines the benefits of both:

■ *Subsumable Learning.* The robot is predisposed to learn certain classes of behaviors by the nature of its existing knowledge. It uses this weak learning ability to adapt to its environment. However, it also has a more general strong learning component, capable of overriding the weaker system when it is forced to learn or adapt to circumstances outside of its predisposition.

2.1   Cost versus Benefit

A question to ask of any learning system proposed or implemented for a robot is: *Is the robot better off for having learning?* In evaluating the answer one must take into account whether investing the same resources of research effort, runtime code size, and computing power into a directly engineered solution would have resulted in an equally robust robot. There has been scant demonstration of robots being better off for having learning. This is not intended as an argument against having learning in general, but rather as an evaluation of progress to date.

Few robot learning systems have demonstrated performance which could not have been built-in or handled by run-time computation. This situation is a result of two different causes. On one hand, infeasible methodologies proliferate since they are not being validated on realistically-sized problems. On the other hand, promising approaches are tested only on the same toy-sized problems, which obscures their potential.

In order for a robot to be adaptive it must be able to learn in a large space of behaviors. However, if the space to be learned is very large, current approaches cannot handle it in reasonable time. On the other hand, if the space is manageable, it is manageable with non-learning solutions as well.

In the next section we divide learnable information into four distinct categories, each of which can be approached with different types of learning methods.

3 THE MAIN TYPES OF LEARNING

Everything a robot learns must eventually be grounded in action in the real world. Specifically, learning has the purpose of facilitating the actions the robot takes, of making them more relevant, appropriate, and precise. The level at which the learned information affects action determines the type of learning methods that will be applicable. In particular, based on the type of information that is learned and its effect on the robot's action in the world, we can divide robot learning approaches into four main categories:

- *Learning numerical functions for calibration or parameter adjustment.* This type of learning optimizes operational parameters in an existing behavioral structure.

- *Learning about the world.* This type of learning constructs and alters some internal representation of the world.

- *Learning to coordinate behaviors.* This type of learning uses existing behavioral structures and their effects in the world to change the conditions and sequences in which they are triggered.

- *Learning new behaviors.* This type of learning builds new behavioral structures.

In the following sections we will examine each of these types of learning in turn.

3.1 Learning Functions

In many cases the numerical parameters of a robot cannot be predicted. Due to sensor drift, environmental conditions, and unmodeled properties of the mechanical system, run-time experience is necessary for choosing or computing these values. Further, physical system parameters are usually coupled, thus making the adjustment problem more complex.

Function learning is the weakest form of learning as the structure of the behavior-producing programs is predetermined and does not change based on experience. Rather than introducing new knowledge, this type of learning tunes the existing behavioral strategy. Based on the number of successfully performing robots, the function approximation approach has been most successful to date.

The two key issues to be addressed in function learning are the shape of the space to be searched, and the nature of the feedback signal used to direct that search.

For high dimensional spaces or complex functions it is necessary to gain experience at many points within the space. The higher the dimension the more experiences are required and the longer it takes to learn a good approximation to the desired function.

With real robots the "correct" function is not known ahead of time. Instead, the closeness of the approximation must be inferred from the performance of the system. [Atkeson 1990a] approaches this problem by recording the system behavior for many trials in different parts of the space and interpolating in order to predict system performance elsewhere, and hence move towards a set of high performance parameters. In another approach [Viola 1990] built a head-eye system with open loop control that learned a transfer function by using retinal slip as the error signal. Although the error metric was too computationally expensive to be used as a direct feedback signal, it was nevertheless used to improve performance.

Classical neural networks are another class of commonly used function learners. They typically operate on very high dimensional spaces, are given correct outputs as teaching examples, and require many trials to cover the search space well enough to get reasonable approximations of the function to be learned.

In contrast to parameter learning, which relies on carefully built-in knowledge, in the following sections we describe methods dealing with less structured types of learning.

3.2 Learning About the World

Most learning research in AI fits into the category of learning about the world. In this approach, some representation of the world is constructed, updated, and refined, based on information gained from the world. Traditionally, the function of the stored information is not known at the time it is stored. Consequently, the information must be represented in some abstract, usually symbolic, form, which is used for computing actions.

Depending on the amount of built-in information, learning about the world can vary from learning maps of the environment to learning abstract concepts [Kuipers 1987]. The latter form is more knowledge-intensive, in that it employs a built-in domain theory in order to minimize the amount of deduction left to the robot, as well as the amount of new information needed from the world [Mitchell *et al.* 1986, Mitchell *et al.* 1989].

Since everything a robot learns must eventually be connected to the way it acts, a more functional approach to learning about the world can be applied to physical robots. Instead of converting the information from the world into an abstract model, and then converting it back into a functional representation, the structure that is learned can be represented as, and grounded in, control action. Such grounded representations are capable of supporting multiple goals and predictions, and can be applied to different types of tasks.

Learning Maps

In this section we describe a system which uses such a functional, grounded representation for encoding new information about the world [Mataric 1992]. The robot's domain was spatial mapping; it explored its environment, recognized new landmarks, and used those to continuously construct and update a map. The user could specify any landmark within

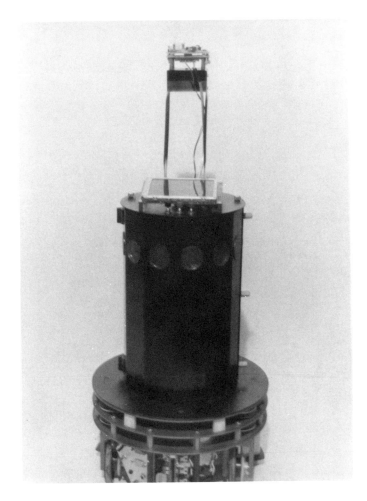

Figure 1. This robot, Toto, learned a procedural representation of the structure of its environment.

the map as the goal, and the robot would execute the shortest path to that goal within the map.

The method was implemented on a mobile robot, Toto, programmed with a basic reflexive behavior of following boundaries of objects in the environment. The representation of the environment consists of a collection of behaviors, each of which corresponds to a unique landmark in the world. Landmarks themselves are defined in terms of behaviors. For example, a right wall is a landmark defined in terms of the robot moving straight and consistently detecting short sonar readings on its right side, and long ones on its left side.

Each of the landmarks stores some descriptive information about itself, including its type (wall, corridor, irregular boundary) and its approximate length. As new landmarks are discovered, a map is constructed by linking landmarks with their topological neighbors. The result is a network isomorphic to the topology of the landmarks in physical space.

Shortest paths within this map are found by spreading activation (weighted by the landmark distance) from the goal throughout the network. Whatever its position, the robot can make a locally optimal choice of direction, based on the local activation, which will result in the globally shortest path to the goal. If a path is blocked, the map is updated by removing the corresponding topological link, and spreading activation through the new network.

This system learns two types of information: 1) the descriptors of the different landmarks it encounters, and 2) the topological relationships between the landmarks. The first type of the information is contained declaratively within the behaviors, filling the built-in slots. In contrast, the topological information is encoded functionally, in terms of what behavior the robot would execute to transition from one landmark to another. While the contents of the robot's behaviors are predetermined, their connections are learned and updated at run-time.

The key to making this system work was selecting the right behavioral primitives for extracting the necessary information from the world [Mataric 1990b]. Further, instead of representing the learned information in a declarative form, it was encoded in the robot's motion behaviors, thus leaving it directly tied to action. In summary, the purpose of learning in this system was not to fundamentally alter the robot's basic behavior, but rather to guide it in the right direction at critical points in its task.

Learning Object Positions

In another application of the same methodology, we implemented a strategy for learning about and copying stacks of pucks. This approach was implemented and tested on a mobile robot equipped with sensors for detecting pucks, and a fork lift for transporting them. The learning was built on top of a basic reactive exploratory and puck searching behavior. Each unique puck was stored in a behavior containing its descriptor, and its position within a stack was represented with the topological links connecting that behavior with other pucks in the network. The topology reflected the stack structure as the nodes were ordered according to the time they were discovered.

To copy a stack, the system "walks up" the representation of the original and finds a matching copy of each puck in order. The representation of the stack implicitly encodes the plan of action; as a puck is copied, activation is passed to the next one in the network. As in the map learning example, the system adds the learned network on top of the existing reflexive behaviors, and uses it to guide and sequence its decisions.

In the two examples we described, the robots learned about the world directly from their actions, and encoded the learned information in the same form in which it originated. This approach suggests that learning behaviors and their sequences can be done without the use of intermediate, abstract, representations.

3.3   Learning to Coordinate Behaviors

Learning to coordinate behaviors attempts to solve the action selection problem, i.e. to determine when particular actions or behaviors are to be executed. Reinforcement learning (RL) methods have been shown to be well suited for this type of learning, as they produce precisely the kind of mapping between conditions and actions needed to decide how to behave at each distinct point in the state space. And indeed, much work has been done on this type of learning, mostly in simulation [Whitehead and Ballard 1990, Sutton 1990, Maes 1991], and some on embodied robots [Maes and Brooks 1990, Mahadevan and Connell 1990].

Learning to Walk

[Maes and Brooks 1990] present a variation of reinforcement learning applied to a simple six-legged robot, Genghis, learning to coordinate leg behaviors. The only reinforcement is negative, based on the state of a switch determining whether the belly of the robot is touching the floor.

Figure 2. This six-legged robot, Genghis, learned to coordinate its leg movements to allow walking.

In previous experiments the robot had been programmed to walk without any learning — all the coordination was hard-wired by the programmer [Brooks 1989].

The learning idea is a generalization of the behavior activation scheme of [Maes 1989]. Independent behaviors on each leg monitor the activity of other behaviors and correlate that activity, their own activity state, and the results from the belly switch, as input to a local learning rule. The correlation is used to learn under which conditions each behavior is to become active. Each behavior builds a logical sentence describing the preconditions of its own activation. For instance, the behavior which lifts up the left front leg learns that it should not be active (i.e. lift up) if the right front leg is already lifted.

After about 20 trials per leg, spread over a total of a minute or two, the robot reliably learns the alternating tripod gait — it slowly emerges out of initially chaotic flailing of the legs. However, the walking gait learned by the robot is not as robust as that programmed by hand.

Although more general simulations were carried out, a bias was built in to the system on the physical robot, because of the small number of behaviors which could be stored on board. The implemented set of behaviors biased the learning procedure to learn to walk forwards, rather then stand still or walk backwards. However the implemented set of behaviors were unbiased with respect to gait selection.

The hardest aspect of generalizing this technique involves the issues of time and credit assignment. The walking task is naturally structured so that only behaviors which are activated simultaneously can lead to negative feedback. One can easily imagine more complex circumstances involving a longer temporal delay between a behavior and its negative consequence. The simple learning algorithm used on Genghis would not be able to cope with such problems.

Delayed reinforcement is only one of the problems involved in using reinforcement learning. In a later section we will discuss other relevant properties of RL as applied to learning in robots.

3.4 Learning New Behaviors

Learning new behaviors requires constructing goal-achieving sequences from primitive actions. The prime motivation for this type of learning is to ease the programmer's job, especially in the case of control tasks which involve intricate interactions among behaviors and among actions within behaviors. Writing a completely new behavior for a robot is a complex task, at least as difficult as writing any other computer program. The latter has not succumbed to machine learning techniques, so we should be skeptical of promises of the former.

Reinforcement learning techniques have been used to learn behaviors [Mahadevan and Connell 1990], but only in the sense that behaviors are arbitrary sequences of actions. RL approaches are well suited to learning such sequences as parts of goal-achieving policies. However, these approaches are unable to abstract the behaviors from the sequences. The result of a reinforcement learning algorithm is a "flat" policy, without any hierarchical properties abstracting behaviors from the state-action pairs.

In order to perform this type of abstraction, some *a priori* structure is necessary. The work of [Mahadevan and Connell 1990] illustrates this well, by showing that a box-pushing behavior is learned more effectively when subdivided into relevant subtasks. However, these subtasks must be determined by the programmer, as they cannot be learned by the system.

An entirely different approach to learning behaviors that holds some promise is the use of genetic programming. This type of learning has been tried in simulation (e.g. [Koza 1990]) but has not yet been successfully applied to real robots. Simulations are an option, but it is very rare that a program run on a simulated robot will transfer to a physical robot. A later section discusses this topic in more detail.

Learning new behaviors has not yet been demonstrated on robots. Although it is perhaps the hardest type of learning, it has not yet received much research attention.

4   THE MAIN METHODS OF LEARNING

It is often said that "one cannot learn anything unless one almost knows it already". The tradeoff between the amount of built-in and learned information is the key issue in learning. While reducing built-in structure eases the programming task and reduces the learning bias, it slows down the learning process, making it less suitable for embodied robots.

We used the type and amount of built-in structure as one of the criteria for categorizing different types of learning. In numerical function learning, the basic behaviors are built-in, and learning optimizes them. In learning to coordinate behaviors, the behaviors are programmed in as well, but the conditions for selecting them are learned. In learning about the world the new information can be used both for selecting and tuning existing behaviors. In the one style of learning which has not yet been demonstrated, whole new behaviors are constructed.

Reinforcement and genetic learning were mentioned as two popular learning methods with applications for real robots. However, both RL and genetic learning require a large number of trials, which may make them ill-suited for embodied robots but appealing for work in simulation. In the following sections we discuss these approaches in detail and address the role of simulations in learning with real robots.

4.1   Reinforcement Learning on Real Robots

Reinforcement learning (RL) has been used for both learning new behaviors and learning to coordinate existing ones. It is currently perhaps the most popular methodology for various types of learning. In this section we review some of its main properties and its suitability for learning on real robots.

The behavior policy for a robot acting in a relatively well understood world can be implemented as a collection of reactive rules, or so called universal plans [Schoppers 1987]. However, for large problems, hardwiring the entire search space is not reasonable. Reinforcement learning systems attempt to learn the policy by attempting all of the actions in all of the available states in order to rank them in the order of appropriateness. Much like universal plans, the reactive policies learned by RL systems are asymptotically complete [Mataric 1991].

The problem of learning the optimal policy can be cast as searching for paths connecting the current state with the goal in the state space. The longer the distance between a state and the goal, the longer it takes to learn the policy. Breaking the problem into modules effectively shortens the distance between the reinforcement signal and the individual actions, but it requires built-in domain information. Intuitively, the amount of built-in structure and the amount of necessary training trials are inversely proportional. However, in real robots, any critical behavior must be learned with a minimal number of trials, since the robot cannot afford to fail repeatedly. In the cases of less critical learning, where the information from the world can be used to improve, tune or optimize an existing behavior, larger numbers of trials are acceptable. Finally, any information which requires hundreds or thousands of time-consuming learning trials ought to be built-in.

An important weakness of RL approaches is their "unstructured" utilization of the input. Since no explicit domain information is used, the entire space of state-action pairs must be explored, but the space grows exponentially with the number of sensors. However, as sensory information is necessary for intelligent behavior, any effective learning algorithm must improve with increased amount of information.

RL work so far has not demonstrated the ability to use previously learned knowledge to speed up the learning of a new policy. Instead, the robot must either ignore the existing policy, or worse, the current policy may be a detriment to learning the next one. However, work continues to be done in this area [Mahadevan 1992].

Finally, RL depends on the ability to perceive the unique state of the robot in order to map it to the appropriate action. Sensor noise and error increase state uncertainty, which further slows down the learning process. Work on active perception has been directed toward this problem [Whitehead and Ballard 1990].

In spite of its weaknesses, reinforcement learning appears to be a promising direction for learning with real robots, in particular because it uses direct information from the real world to improve the robot's performance.

4.2   Genetic Learning

In this section we return to the problem of learning new behaviors and consider some of the issues in using genetic learning for robot programming.

In nature, evolution contemporaneously experiments with the morphology of the individual and its neural circuitry using the same genetic mechanism. There are two important consequences of this fact, both of which reduce the space which evolution must search.

- The control program is evolved incrementally. Evolution is restricted initially to a small search space. The size of the space grows over many generations, but by then a good partial solution is already found and is used as the basis for searching the newer parts of the space.

■   Symmetric or repeated structures naturally have symmetric or repeated neural circuits installed — they do not need to be individually evolved. (For example, the bodies of *Drosophila* are segmented,
and mutants can be produced with extra segments. Each added
segment contains a leg pair and comes with appropriate neural circuitry. Further, a single encoding specifies the wiring of both a right
and a left leg.)

Our approach to behavior-based programming of robots has always been
to build layers incrementally [Brooks 1986]. For genetic programming
we suggest that the robot (both simulated and physical) should initially
be operated with only some of its sensors, and perhaps only some of its
actuators, to simplify the problem. Programs can be evolved for this simplified robot, much as hand-written programs are [Brooks 1990b]. Once
the fundamental behaviors are present, additional sensors and actuators
can be made available so that higher level behaviors can be evolved. The
particular fitness function used to control the evolutionary search can be
varied over time to emphasize the use of new capabilities and the need to
develop higher level behaviors. There might also be some advantage to
biasing the genetic operators more toward the newer behaviors than the
older ones. At the same time, it certainly makes sense for the crossover
operator to use pieces of an old behavior. For example, an orienting
behavior, based on a sensor activated early in the evolutionary search,
would be a good prototype for an orienting behavior using a new sensor.
To carry out this approach the program must be annotated with a layer
number throughout. One-way crossover might be used rather than two-
way crossover, and it could be biased to alter the structure of the most
recent layers.

In hand-written programs we capture symmetry and repeated structure
through the use of programming constructs like macros. For example, on
the six-legged robot Genghis, a macro version of the leg control behaviors
gets instantiated six times, once for each leg [Brooks 1989]. This suggests
two ideas for reducing the genetic search space:

■   The language that is subject to genetic programming should include
a macro capability. When the search learns how to use this capability, it will greatly accelerate the production of effective programs.

- There must be some way to reference the symmetries and regularities of the morphological structure of the robot, in order to invoke the macros correctly. This requires that the programmer provide an explicit description of the regularities in the morphology of the robot.

These techniques may make genetic programming more tractable as a way of automatically generating completely new behaviors. If genetic techniques are to be pursued for learning new behaviors, then we must address the issues of having function follow morphology, and of controlling search through layering the robot's capabilities. Most importantly, we must find a way of evaluating genetic programs in such a way that the results are valid on physical robots.

5 SIMULATION

There is great appeal to using simulated robots for investigating robot learning. Besides being affordable, simulations simplify the logistics of experimentation. However, relying on simulations can be dangerous as they can inadvertently lead to the development and investigation of algorithms which are not transferable to real robots. This is not to say that the algorithms may not be of interest in some domains — merely that they may be hard or impossible to transfer to real robot domains.

5.1 Simulating the World

Simulations are doomed to succeed. Even despite best intentions there is a temptation to fix problems by tweaking the details of a simulation rather than the control program or the learning algorithm. Another common pitfall is the use of global information that could not possibly be available to a real robot. Further, it is often difficult to separate the issues which are intrinsic to the simulation environment from those which are based in the actual learning problem.

Simulations cannot be made sufficiently realistic. One of the problems with simulated worlds which specifically affects the success of learning algorithms, is their assumed determinism. Some learning strategies allow for particular limited kinds of error and noise in the world, usually in the form of specific percentage of randomness in the input or the reinforcement. However, unlike simulations, the real world is not necessarily deterministic at a modelable level, and its nondeterminism usually does not fit a simple stochastic model. For example, stochastic models do not capture the non-regularity due to rapidly changing boundary conditions caused by motion. Sensor readings change quite abruptly as real and illusory object appear and disappear.

Many simulation artifacts do not occur in the real world. In particular, many simulations are concerned with resolving various oscillations and symmetries between robots. The dynamics actually tend to be more brittle than in the real world where noise and stochastic processes have a smoothing effect. Thus, while simulated worlds are in many ways simpler than the real world, they are paradoxically sometimes harder to work with. For example, much research has been concerned with path planning for mobile robots in non-grid worlds, i.e. in two dimensional Euclidean space. Some of this work expends much effort on solving the problem of the paths of two robots crossing each other and introduces elaborate protocols to avoid deadlock. But real robots never reach the state of perfect deadlock. They never run down their respective corridors perfectly and arrive at identical times. Simple reactive strategies suffice to break any possible deadlock, just as random variations in Ethernet controllers break deadlocks on rebroadcast. In general, no perfect symmetry or simultaneity is possible in the real world.

5.2   Simulating Sensors and Actuators

Although apparently simple, the following properties are often misunderstood in simulations:

- Sensors deliver very uncertain values even in a stable world.

- The data delivered by sensors are not direct descriptions of the world. They do not directly provide high level object descriptions and their relationships.

- Commands to actuators have very uncertain effects.

It is difficult to interpret sensor data in the real world. Even carefully calibrated sensors can produce unexpected readings when introduced into a complex dynamic environment due to effects such as material properties, multiple reflections, specularities, etc. In addition to these systematic interactions, sensors tend to be highly susceptible to random noise. These problems are common to all types of sensors, including vision, ultrasound, and infra-red. At best, sensors deliver a fuzzy approximation to what they appear to be measuring.

Sensor readings are not complete descriptions of the world. For example, many simulations assume the existence of a sensor which provides information such as *is there food ahead?* Real sensors do not deliver such high level information but rather measure indirect aspects of the world, such as distance or albedo. They do not separate objects from the background. They do not identify objects. They do not give pose information about objects. They do not separate out static objects, moving objects, and effects due to self motion. They do not operate in a stable coordinate system independent of the uncertain motion of the robot. They do not integrate data from multiple sensors into a single consistent picture of the world.

In order to provide the high level information assumed in simulations, a robot would require a complex perceptual system. Such systems do not exist, so that even working simulations do not transfer to physical robots. Further, it may be impossible to treat perception as a black box with a clean interface to the rest of intelligence [Brooks 1991b, Brooks 1991a].

Just as sensors do not deliver simple descriptions of the world, high level action commands need many layers of refinement before they become appropriately orchestrated motor currents. For example, a high level action such as *open door* cannot be thought of as a primitive command to an actuator.

In physical robots, sensing and action are intimately connected. They both generate and at the same time rely on the dynamics of the robot's interaction with the world. Simple state space models of the world, and black box models of perception are not sufficiently realistic to make simulations which employ them useful. Simulations can be a beneficial tool for testing algorithms but robot learning algorithms must ultimately be validated on physical robots.

6 CONCLUSION

In this paper we have examined the problem of learning on embodied robots. We proposed a division of the various learning styles into four main categories, based on the amount of built-in information and the type of information being learned. Of these, most successful work has been done in the area of parameter learning.

Some successful demonstrations of learning about the world have been shown on real robots. The traditional approach of learning disembodied symbolic descriptions of the world, however, has not been successfully applied to embodied robots. Many robots have used symbolic descriptions of the world, but they have been pre-supplied descriptions, or lead through a training phase by an external teacher. Other robots have used local maps of the world based on a purely geometric description. In that case, detailed traversabilty knowledge of geometric configurations has been built into the robot's control program. We described an alternative approach, which grounds the descriptions of the world directly in the control of the robot's actions.

For the problem of coordinating behaviors, variations of reinforcement learning have been used in a small number of successful real robot applications. However, some basic properties of RL make it ill-suited for larger scale learning in real-time.

Finally, learning new behaviors has not yet been demonstrated. This type of learning may be a particularly difficult goal to achieve — some ethological evidence suggests that most animal species are not capable of it [Gould 1982]. We discussed the application of genetic learning to this problem.

The weaknesses of the existing learning techniques, and the versatility of the knowledge necessary to make a robot perform efficiently in the real world suggest that many concurrent, complementary, and redundant learning methods may be necessary.

> *Learning about the real world with real sensors on an embodied system is hard. Real robots need to learn many qualitatively different types of information and they will require many different methods to learn each of them.*

Acknowledgements

We are grateful to Michael Bolotski, Nancy Pollard, and the editors, whose helpful comments and suggestions greatly improved this paper.

This report describes research done at the Artificial Intelligence Laboratory of the Massachusetts Institute of Technology. Support for this research was provided in part by the Advanced Research Projects Agency under Office of Naval Research contract N00014-85-K-0124, in part by the Hughes Artificial Intelligence Center, and in part by Mazda Corporation.

BIBLIOGRAPHY

[Aboaf et al. 1988] Eric W. Aboaf, Christopher G. Atkeson, and David J. Reinkensmeyer. Task-level robot learning. In *Proceedings of the 1988 IEEE International Conference on Robotics and Automation*, pages 1309–1310, 1988.

[Aboaf 1988] Eric W. Aboaf. Task-level robot learning. Technical Report AI TR 1079, Massachusetts Institute of Technology, 1988.

[Albus 1981] James Albus. A neurological model. In *Brains, Behavior, and Robotics*. Byte Books, Peterborough NH, 1981.

[Allen and Roberts 1989] Peter K. Allen and Kenneth S. Roberts. Haptic object recognition using a multi-fingered dextrous hand. In *Proceedings of the 1989 IEEE International Conference on Robotics and Automation*, pages 342–347, 1989.

[Andreae 1985] Peter Merrett Andreae. Justified generalization: Acquiring procedures from examples. Technical Report AI TR 834, Massachusetts Institute of Technology, 1985.

[Angluin 1978] Dana Angluin. On the complexity of minimum inference of regular sets. *Information and Control*, 39:337–350, 1978.

[Angluin 1987] Dana Angluin. Learning regular sets from queries and counterexamples. *Information and Computation*, 75:87–106, 1987.

[Arbib and House 1987] Michael A. Arbib and Donald H. House. Depth and detours: An essay on visually guided behavior. In M. Arbib and A. Hanson, editors, *Vision, Brain, and Cooperative Computation*. MIT Press, Cambridge, MA, 1987.

[Asada and Yang 1989] Haruhiko Asada and Boo-Ho Yang. Skill acquisition from human experts through pattern processing of teaching data. In *Proceedings of the 1989 IEEE International Conference on Robotics and Automation*, pages 1302–1307, 1989.

215

[Asada 1990] Haruhiko Asada. Teaching and learning of compliance using neural nets: Representation and generation of nonlinear compliance. In *Proceedings of the 1990 IEEE International Conference on Robotics and Automation*, pages 1237–1244, 1990.

[Atkeson 1989] Christopher G. Atkeson. Learning arm kinematics and dynamics. *Annual Review of Neuroscience*, 12:157–183, 1989.

[Atkeson 1990a] Christopher G. Atkeson. Memory-based approaches to approximating continuous functions. In *The Sixth Yale Workshop on Adaptive and Learning Systems*, August 1990.

[Atkeson 1990b] Christopher G. Atkeson. Using associative content-addressable memories to control robots. In Patrick Henry Winston and Sarah Alexandra Shellard, editors, *Artificial Intelligence at MIT: Expanding Frontiers, Volume 2*. MIT Press, Cambridge MA, 1990.

[Ballard and Brown 1982] Dana H. Ballard and Christopher M. Brown. *Computer Vision*. Prentice Hall, Englewood Cliffs NJ, 1982.

[Barto et al. 1983] A. G. Barto, R. S. Sutton, and C. W. Anderson. Neuronlike Adaptive elements that that can learn difficult Control Problems. *IEEE Trans. on Systems Man and Cybernetics*, 13(5):835–846, 1983.

[Barto et al. 1989] A. G. Barto, R. S. Sutton, and C. J. C. H. Watkins. Learning and Sequential Decision Making. COINS Technical Report 89-95, University of Massachusetts at Amherst, September 1989.

[Barto et al. 1991] A. G. Barto, S. J. Bradtke, and S. P. Singh. Real-time Learning and Control using Asynchronous Dynamic Programming. Technical Report 91-57, University of Massachusetts at Amherst, August 1991.

[Barto 1987] Andrew G. Barto. An approach to learning control surfaces by connectionist systems. In M. Arbib and A. Hanson, editors, *Vision, Brain, and Cooperative Computation*. MIT Press, Cambridge MA, 1987.

[Basye et al. 1989] Kenneth Basye, Thomas Dean, and Jeffrey Scott Vitter. Coping with uncertainty in map learning. In *Proceedings IJCAI 11*, pages 663–668, 1989.

[Bellman 1957] R. E. Bellman. *Dynamic Programming*. Princeton University Press, Princeton, NJ, 1957.

[Berry and Fristedt 1985] D. A. Berry and B. Fristedt. *Bandit Problems: Sequential Allocation of Experiments*. Chapman and Hall, 1985.

[Bertsekas and Tsitsiklis 1989] D. P. Bertsekas and J. N. Tsitsiklis. *Parallel and Distributed Computation*. Prentice Hall, 1989.

[Bertsekas 1987] D. P. Bertsekas. *Dynamic Programming: Deterministic and Stochastic Models*. Prentice-Hall, 1987.

[Blaauw 1982] G.J Blaauw. Driving experience and task demands in simulator and instrumented car: A validation study. *Human Factors*, 24:473–486, 1982.

[Bolle *et al.* 1989] Ruud M. Bolle, Andrea Califano, Rick Kjeldsen, and Russell W. Taylor. Visual recognition using concurrent and layered parameter transforms. In *Proccedings of the IEEE Conference on Computer Vision and Pattern Recognition*, pages 625–631, June 1989.

[Borenstein and Koren 1989] Johann Borenstein and Yorem Koren. Real-time obstacle avoidance for fast mobile robots. *IEEE Transactions on Systems, Man, and Cybernetics*, 19:1179–1187, October 1989.

[Borenstein and Koren 1991] Johann Borenstein and Yorem Koren. The vector field histogram – fast obstacle avoidance for mobile robots. *IEEE Journal of Robotics and Automation*, 7:278–288, June 1991.

[Breiman *et al.* 1984] Leo Breiman, Jerome H. Friedman, Richard A Olshen, and Charles J. Stone. *Classification and Regression Trees*. Wadsworth and Brooks, Monterrey CA, 1984.

[Brooks 1986] Rodney A. Brooks. A robust layered control system for a mobile robot. *IEEE Journal of Robotics and Automation*, RA-2(1):14–23, March 1986.

[Brooks 1989] Rodney A. Brooks. A robot that walks: Emergent behavior from a carefully evolved network. *Neural Computation*, 1(2):253–262, 1989.

[Brooks 1990a] Rodney A. Brooks. The behavior language; user's guide. AI Memo 1227, MIT Artificial Intelligence Lab, Cambridge, MA, 1990.

[Brooks 1990b] Rodney A. Brooks. Elephants don't play chess. In Pattie Maes, editor, *Designing Autonomous Agents*, pages 3–15. Elsevier/MIT Press, Cambridge, MA, 1990.

[Brooks 1991a] Rodney A. Brooks. Intelligence without reason. In *Proceedings of IJCAI-91*, August 1991. (Also MIT AI Memo 1293, MIT Artificial Intelligence Laboratory, April 1991).

[Brooks 1991b] Rodney A. Brooks. Intelligence without representation. *Artificial Intelligence*, 47:139–160, 1991.

[Chapman and Kaelbling 1991] David Chapman and Leslie Pack Kaelbling. Learning from delayed reinforcement in a complex domain. In *Proceedings of IJCAI*, 1991. (Also Teleos Technical Report TR-90-11, 1990).

[Chatila and Laumond 1985] Raja Chatila and Jean-Paul Laumond. Position referencing and consistent world modeling for mobile robots. In *Proceedings IEEE International Conference on Robotics and Automation*, pages 138–170, 1985.

[Chrisman 1992] Lonnie Chrisman. Reinforcement learning with perceptual aliasing. In *Proceedings of the Eleventh National Conference on Artificial Intelligence*, 1992.

[Clouse and Utgoff 1992] Jeffery Clouse and Paul Utgoff. A teaching method for reinforcement learning. In *Proceedings of the Ninth International Conference on Machine Learning*, pages 92–101. Morgan Kaufmann, 1992.

[Connell and Brady 1987] Jonathan H. Connell and J. Michael Brady. Generating and generalizing models of visual objects. *Artificial Intelligence*, 31:159–183, 1987.

[Connell 1989] Jonathan H. Connell. A behavior-based arm controller. *IEEE Transactions on Robotics and Automation*, 5(6):784–791, 1989.

[Connell 1990] Jonathan H. Connell. *Minimalist Mobile Robotics: A Colony-style Architecture for an Artificial Creature*. Academic Press, San Diego, CA, 1990. (revised version of MIT Ph.D. thesis, August 1989).

[Connell 1991] Jonathan H. Connell. Controlling a mobile robot using partial representations. In *Proceedings of the 1991 SPIE Conference on Mobile Robots*, pages 34–45, 1991.

[Connell 1992] Jonathan H. Connell. SSS: A hybrid architecture applied to robot navigation. In *Proceedings of the 1992 IEEE International Conference on Robotics and Automation*, pages 2719–2724, 1992.

[Cottrell 1990] G.W Cottrell. Extracting features from faces using compression networks: Face, identity, emotion and gender recognition using holons. In David Touretzky, editor, *Connectionist Models: Proc. of the 1990 Summer School*, San Mateo, CA, 1990. Morgan Kaufmann.

[Crisman and Thorpe 1990] J.D. Crisman and C.E. Thorpe. Color vision for road following. In Charles Thorpe, editor, *Vision and Navigation: The CMU Navlab*. Kluwer Academic Publishers, Boston, MA, 1990.

[Crowley *et al.* 1991] James L. Crowley, Philippe Bobet, Karen Sarachik, Stephen Mely, and Michel Kurek. Mobile robot perception using vertical line stereo. *Journal of Robotics and Autonomous Systems*, 7:125–138, 1991.

[Crowley 1985] James L. Crowley. Dynamic world modeling for an intelligent mobile robot using a rotating ultra-sonic ranging device. In *Proceedings IEEE International Conference on Robotics and Automation*, pages 128–135, 1985.

[Dayan 1992] P. Dayan. The Convergence of TD(λ) for General λ. *Machine Learning*, 8(3), May 1992.

[Dean *et al.* 1990] Thomas Dean, Kenneth Basye, Robert Chekaluk, Seungseok Hyun, Moises Lejter, and Margaret Randazza. Coping with uncertainty in a control system for navigation and exploration. In *Proceedings AAAI-90*, pages 1010–1015, 1990.

[Dean *et al.* 1992] Thomas Dean, Dana Angluin, Kenneth Basye, Sean Engelson, Leslie Kaelbling, Evangelos Kokkevis, and Oded Maron. Inferring finite automata with stochastic output functions and an application to map learning. In *Proceedings AAAI-92*, pages 208–214, 1992.

[Dickmanns and Zapp 1987] E.D. Dickmanns and A. Zapp. Autonomous high speed road vehicle guidance by computer vision. In *Proceedings of the 10th World Congress on Automatic Control, Vol. 4*, 1987.

[Donald and Xavier 1989] Bruce Donald and Patrick Xavier. A provably good approximation algorithm for optimal-time trajectory planning. In *Proceedings IEEE International Conference on Robotics and Automation*, pages 958–963, 1989.

[Drescher 1991] Gary L. Drescher. *Made-up Minds-A Constructivist Approach to Artificial Intelligence*. MIT Press, Cambridge, MA, 1991.

[Drumheller 1987] Michael Drumheller. Mobile robot localization using sonar. *IEEE Trans. on Pattern Analysis and Machine Intelligence*, PAMI-9(2):325–332, 1987.

[Duda and Hart 1973] R. O. Duda and P. E. Hart. *Pattern Recognition and Scene Analysis*. Wiley, New York NY, 1973.

[Dudek et al. 1991] Gregory Dudek, Michael Jenkin, Evangelos Milios, and David Wilkes. Robotic exploration as graph construction. *IEEE Transactions on Robotics and Automation*, 7(6):859–865, 1991. (Also Technical Report RBCV-TR-88-23, University of Toronto, 1988).

[Dufay and Latombe 1984] Bruno Dufay and Jean-Claude Latombe. An approach to automatic robot programming based on inductive learning. *International Journal of Robotics Research*, 3(4):3–20, 1984.

[Elfes 1987] Alberto Elfes. Sonar based real-world mapping and navigation. *IEEE Journal of Robotics and Automation*, 3(3):249–265, 1987.

[Engel 1989] Antonie J. Engel. Image features as virtual beacons for local navigation. In *Proceedings SPIE Vol. 1002 Intelligent Robots and Computer Vision*, pages 626–633, 1989.

[Engelson and McDermott 1992] Sean Engelson and Drew McDermott. Passive robot map building with exploration scripts. Technical Report Computer Science Department 898, Yale University, 1992.

[Feng and Krogh 1990] Dai Feng and Bruce H. Krogh. Satisficing feedback strategies for local navigation of autonomous mobile robots. *IEEE Transactions on Systems, Man, and Cybernetics*, 20:1383–1395, November 1990.

[Flynn 1985] Anita M. Flynn. Redundant sensors for mobile robot navigation. Tech. Report AI–TR 859, MIT, 1985.

[Foulser *et al.* 1990] D.E. Foulser, M. Li, and Q. Yang. Theory and algorithms for plan merging. Technical Report cs-90-40, University of Waterloo, Computer Science, 1990.

[Friedman *et al.* 1977] Jerome H. Friedman, Jon Louis Bentley, and Raphael Ari Finkel. An algorithm for finding best matches in logarithmic expected time. *ACM Transactions on Mathematical Software*, 3(3):209–226, 1977.

[Froom 1991] Richard Froom. Acquiring effective knowledge of environmental geometry for minimum-time control of a mobile robot. In *Proceedings IEEE International Symposium on Intelligent Control*, pages 501–506, Arlington, VA, August 13 15 1991.

[Fu 1970] K. S. Fu. Learning Control Systems—Review and Outlook. *IEEE Trans. on Automatic Control*, pages 210–221, April 1970.

[Gat and Miller 1990] Erann Gat and David P. Miller. BDL: A language for programming reactive robotic control systems. Tech. report, California Institute of Technology/Jet Propulsion Laboratory, June 1 1990.

[Gilbert and Johnson 1985] Elmer G. Gilbert and Daniel W. Johnson. Distance functions and their application to robot path planning in the presence of obstacles. *IEEE Journal of Robotics and Automation*, RA-1:21–30, March 1985.

[Gold 1972] E. Mark Gold. System identification via state characterization. *Automatica*, 8:621–636, 1972.

[Gould 1982] James L. Gould. *Ethology – The Mechanisms and Evolution of Behavior*. W. W. Norton & Co., New York, 1982.

[Hampshire and Waibel 1989] J.B. Hampshire and A.H. Waibel. The meta-pi network: Building distributed knowledge representations for robust pattern recognition. Technical Report CMU-CS-89-166-R, Carnegie Mellon University, August 1989.

[Holland *et al.* 1986] John H. Holland, Keith J. Holyoak, Richard E. Nisbett, and Paul R. Thagard. *Induction*. MIT Press, Cambridge MA, 1986.

[Holland 1986] John H. Holland. Escaping brittleness: the possibilities of general-purpose learning algorithms applied to parallel rule-based systems. In *Machine Learning: An Artificial Intelligence Approach. Volume II.* Morgan Kaufmann, San Mateo, CA, 1986.

[Jacobs *et al.* 1990] R.A. Jacobs, M.I. Jordan, and A.G. Barto. Task decomposition through competition in a modular connectionist architecture: The what and where vision tasks. Technical Report Computer and Information Science Technical Report 90-27, University of Massachusetts, March 1990.

[Kadonoff *et al.* 1986] Mark B. Kadonoff, Faycal Benayad-Cheriff, Austin Franklin, James F. Maddox, Lon Muller, and Hans Moravec. Arbitration of multiple control strategies for mobile robots. In *Proceedings SPIE Vol. 727 Mobile Robots*, pages 90–98, 1986.

[Kaelbling *et al.* 1992] Leslie Kaelbling, Kenneth Basye, Thomas Dean, Evangelos Kokkevis, and Oded Maron. Robot map-learning as learning labeled graphs from noisy data. Technical Report CS-92-15, Brown University, 1992.

[Kaelbling 1990] L. P. Kaelbling. Learning in Embedded Systems. PhD. Thesis; Technical Report No. TR-90-04, Stanford University, Department of Computer Science, June 1990.

[Khatib 1986] O. Khatib. Real-time obstacle avoidance for manipulators and mobile robots. *International Journal of Robotics Research*, 5(1), 1986.

[Kluge and Thorpe 1990] K. Kluge and C.E. Thorpe. Explicit models for robot road following. In Charles Thorpe, editor, *Vision and Navigation: The CMU Navlab.* Kluwer Academic Publishers, Boston, MA, 1990.

[Knuth 1973] D. E. Knuth. *Sorting and Searching.* Addison Wesley, 1973.

[Korf 1990] R. E. Korf. Real-Time Heuristic Search. *Artificial Intelligence*, 42, 1990.

[Koza 1990] John R. Koza. Evolution and co-evolution of computer programs to control independently-acting agents. In J. Meyer and S. Wilson, editors, *From Animals to Animats: Proceedings of the First*

International Conference on Simulation of Adaptive Behavior, pages 366–375, Cambridge, MA, 1990. MIT Press/Bradford Books.

[Krzanowski 1988] W. J. Krzanowski. *Principles of Multivariate Analysis: A User's Perspective*. Oxford Statistical Science Series. Clarendon Press, Oxford, 1988.

[Kuipers and Byun 1988] Benjamin J. Kuipers and Yung-Tai Byun. A robust, qualitative method for robot spatial learning. In *Proceedings of the National Conference on Artificial Intelligence (AAAI-88)*, pages 774–779, St. Paul/Minneapolis, 1988.

[Kuipers and Byun 1991] Benjamin J. Kuipers and Yung-Tai Byun. A robot exploration and mapping strategy based on a semantic hierarchy of spatial representations. *Journal of Robotics and Autonomous Systems*, 8.47–03, 1991.

[Kuipers and Levitt 1988] Benjamin J. Kuipers and Tod S. Levitt. Navigation and mapping in large-scale space. *AI Magazine*, 9(2):25–43, 1988.

[Kuipers 1977] Benjamin J. Kuipers. Representing knowledge of large-scale space. Tech. Report TR-418, MIT Artificial Intelligence Laboratory, Cambridge, MA, July 1977. (Doctoral thesis, MIT Mathematics Department).

[Kuipers 1978] Benjamin Kuipers. Modeling spatial knowledge. *Cognitive Science*, 2:129–153, 1978.

[Kuipers 1979a] Benjamin J. Kuipers. Commonsense knowledge of space: Learning from experience. In *Proceedings of the Sixth International Joint Conference on Artificial Intelligence*, Stanford, CA, 1979. Stanford Computer Science Department.

[Kuipers 1979b] Benjamin J. Kuipers. On representing commonsense knowledge. In N. V. Findler, editor, *Associative Networks: The Representation and Use of Knowledge by Computers*. Academic Press, New York, 1979.

[Kuipers 1982] Benjamin J. Kuipers. The "map in the head" metaphor. *Environment and Behavior*, 14(2):202–220, 1982.

[Kuipers 1983a] Benjamin J. Kuipers. The cognitive map: Could it have been any other way? In Jr. H. L. Pick and L. P. Acredolo, editors, *Spatial Orientation: Theory, Research, and Application*. Plenum Press, New York, 1983.

[Kuipers 1983b] Benjamin J. Kuipers. Modeling human knowledge of routes: Partial knowledge and individual variation. In *Proceedings of the National Conference on Artificial Intelligence (AAAI-83)*, Los Altos, CA, 1983. Morgan Kaufmann.

[Kuipers 1985] Benjamin J. Kuipers. The map-learning critter. Tech. Report AI TR 85-17, AI Laboratory, University of Texas at Austin, Austin, Texas, 1985.

[Kuipers 1987] Benjamin Kuipers. A qualitative approach to robot exploration and map learning. In *AAAI Workshop on Spatial Reasoning and Multi-Sensor Fusion*, October 1987.

[LeCun *et al.* 1989] Y. LeCun, B. Boser, J.S. Denker, D. Henderson, R.E. Howard, W. Hubbard, and L.D. Jackel. Backpropagation applied to handwritten zip code recognition. *Neural Computation*, 1(4), 1989.

[Lee 1991] Wan-Yik Lee. SIMRX – a simulator for robot exploration, map learning and navigation. A user's manual. (unpublished manuscript), 1991.

[Leonard *et al.* 1990] John Leonard, Hugh Durrant-Whyte, and Ingemar J. Cox. Dynamic map building for an autonomous mobile robot. In *Proceedings IEEE/RSJ International Workshop on Intelligent Robots and Systems*, pages 89–95, 1990.

[Levinson *et al.* 1983] S. E. Levinson, L. R. Rabiner, and M. M. Sondhi. An introduction to the application of the theory of probabilistic functions of a Markov process to automatic speech recognition. *The Bell System Technical Journal*, 62(4):1035–1074, 1983.

[Levitt and Lawton 1990] Tod S. Levitt and Daryl T. Lawton. Qualitative navigation for mobile robots. *Artificial Intelligence*, 44:303–360, 1990.

[Levitt *et al.* 1987] Tod S. Levitt, Daryl T. Lawton, David M. Chelberg, and Philip C. Nelson. Qualitative landmark-based path planning and following. In *Proceedings AAAI-87*, pages 689–694, 1987.

[Lin 1991a] Long Ji Lin. Self-improvement based on reinforcement learning, planning, and teaching. In *Proceedings of the Eighth International Workshop on Machine Learning*, 1991.

[Lin 1991b] Long-Ji Lin. Programming Robots using Reinforcement Learning and Teaching. In *Proceedings of the ninth International conference on Artificial Intelligence (AAAI-91)*. MIT Press, 1991.

[Lippmann 1987] Richard P. Lippmann. An introduction to computing with neural networks. *IEEE Acoustics, Speech, and Signal Processing Magazine*, pages 4–22, April 1987.

[Lynch 1960] Kevin Lynch. *The Image of the City*. MIT Press, Cambridge, MA, 1960.

[Maes and Brooks 1990] Pattie Maes and Rodney A. Brooks. Learning to coordinate behaviors. In *Proceedings of AAAI-90*, pages 796–802, 1990.

[Maes 1989] Pattie Maes. The dynamics of action selection. In *Proceedings of IJCAI-89*, pages 991–997, 1989.

[Maes 1991] Pattie Maes. Learning behavior networks from experience. In *Proceedings of the European A-Life Conference*, 1991.

[Mahadevan and Connell 1990] Sridhar Mahadevan and Jonathan H. Connell. Automatic programming of behavior-based robots using reinforcement learning. Technical Report RC 16359, IBM T. J. Watson Research Center Research, Yorktown Heights, NY, December 1990.

[Mahadevan and Connell 1991] Sridhar Mahadevan and Jonathan H. Connell. Scaling reinforcement learning to robotics by exploiting the subsumption architecture. In *Proceedings of the Eighth International Workshop on Machine Learning*, 1991.

[Mahadevan and Connell 1992] Sridhar Mahadevan and Jonathan H. Connell. Automatic programming of behavior-based robots using reinforcement learning. *Artificial Intelligence*, 55:311–365, 1992. (a revised version of IBM Tech Report RC 16359).

[Mahadevan 1992] Sridhar Mahadevan. Enhancing transfer in reinforcement learning by building stochastic models of robot actions. In *Proceedings of the Ninth International Conference on Machine Learning*, pages 290–299, 1992.

[Manteuffel 1991] Gerhard Manteuffel. A biological visuo-motor system: how dissimilar maps interact to produce behavior. In *From Animals to Animats*, pages 120–126. MIT Press, Cambridge MA, 1991. (Proceedings of SAB-90).

[Mataric 1990a] Maja J. Mataric. A distributed model for mobile robot environment-learning and navigation. Technical Report TR-1128, MIT AI Laboratory, 1990.

[Mataric 1990b] Maja J. Mataric. Navigating with a rat brain: A neurobiologically-inspired model for robot spatial representation. In J. Meyer and S. Wilson, editors, *From Animals to Animats: Proceedings of the First International Conference on Simulation of Adaptive Behavior*, pages 169–175, Cambridge, MA, 1990. MIT Press/Bradford Books.

[Mataric 1991] Maja J. Mataric. A comparative analysis of reinforcement learning methods. Technical Report Memo 1322, MIT AI Lab, Cambridge, MA, October 1991.

[Mataric 1992] Maja J. Mataric. Integration of representation into goal-driven behavior-based robots. *IEEE Transactions on Robotics and Automation*, 8(3), June 1992.

[Mel 1990] Bartlet W. Mel. *Connectionist Robot Motion Planning: A Neurally-Inspired Approach to Visually-Guided Reaching*. Academic Press, San Diego, CA, 1990, (revised version of University of Illinois Ph.D. thesis, Report CCSR-89-17A, March 1989).

[Mel 1991] Bartlet W. Mel. Further explorations in visually-guided reaching: Making MURPHY smarter. In *IEEE Conference on Neural Information Processing Systems – Natural and Synthetic*, pages 348–355, 1991.

[Michie and Chambers 1968] D. Michie and R. A. Chambers. BOXES: An Experiment in Adaptive Control. In *Machine Intelligence 2*. Oliver and Boyd, 1968.

[Mihail 1989] Milena Mihail. Conductance and convergence of Markov chains: A combinatorial treatment of expanders. In *Proceedings of the Thirty First Annual Symposium on Foundations of Computer Science*, pages 526–531, 1989.

[Miller III *et al.* 1987] Thomas W. Miller III, Filson H. Glanz, and L. Gordon Kraft III. Application of a general learning algorithm to the control of robotic manipulators. *International Journal of Robotics Research*, 6(2):84–98, 1987.

[Miller III 1986] Thomas W. Miller III. A nonlinear learning controller for robotic manipulators. In *Proceedings of the 1986 SPIE Conference on Mobile Robots*, pages 416–423, 1986.

[Minsky and Papert 1969] Marvin L. Minsky and Seymour A. Papert. *Perceptrons*. MIT Press, Cambridge MA, 1969.

[Mitchell and Thrun 1992] Tom M. Mitchell and Sebastian B. Thrun. Explanation-based neural network learning for robot control. In C. L. Giles, S. J. Hanson, and J. D. Cowan, editors, *Advances in Neural Information Processing Systems 5*. Morgan Kaufmann, San Mateo CA, 1992.

[Mitchell *et al.* 1986] Tom M. Mitchell, Richard M. Keller, and Smadar T. Kedar-Cabelli. Explanation-based generalization: A unifying view. *Machine Learning*, 1:47–80, 1986.

[Mitchell *et al.* 1989] Tom M. Mitchell, Matthew T. Mason, and Alan D. Christiansen. Toward a learning robot. Technical Report CS-89-106, Carnegie Mellon University, Pittsburgh, PA, January 1989.

[Mitchell 1990] Tom M. Mitchell. Becoming increasingly reactive. In *Proceedings of the Eighth AAAI*, pages 1051–1058. Morgan Kaufmann, 1990.

[Mitchie and Chambers 1968] D. Mitchie and R. A. Chambers. BOXES: An experiment in adaptive control. In E. Dale and D. Mitchie, editors, *Machine Intelligence 2*. Oliver and Boyd, Edinburgh, Scotland, 1968.

[Moore and Atkeson 1992] A. W. Moore and C. G. Atkeson. Memory-based Function Approximators for Learning Control. In preparation, 1992.

[Moore and Golledge 1976] G. T. Moore and R. G. Golledge, editors. *Environmental Knowing: Theories, Research, and Methods*. Dowden, Hutchinson and Ross, Stroudsburg, PA, 1976.

[Moore 1990] Andrew Willima Moore. *Efficient Memory-based Learning for Robot Control*. PhD thesis, Cambridge University, October 1990.

[Moore 1991] A. W. Moore. Variable Resolution Dynamic Programming: Efficiently Learning Action Maps in Multivariate Real-valued State-spaces. In *Machine Learning: Proceedings of the Eighth International Workshop*. Morgan Kaufman, June 1991.

[Moravec and Elfes 1985] H. Moravec and A. Elfes. High resolution maps from wide angle sonar. In *IEEE Robotics and Automation*, pages 116–121, 1985.

[Moravec 1988] Hans Moravec. Certainty grids for mobile robots. *AI Magazine*, 9(2):61–74, 1988.

[Newell and Simon 1976] Allen Newell and Herbert A. Simon. Computer science as empirical inquiry. *Communications of the Association for Computing Machinery*, 19:113–126, 1976.

[Nilsson 1971] N. J. Nilsson. *Problem-solving Methods in Artificial Intelligence*. McGraw Hill, 1971.

[Peng and Williams 1992] J. Peng and R. J. Williams. Efficient Search Control in Dyna. College of Computer Science, Northeastern University, March 1992.

[Piaget and Inhelder 1967] Jean Piaget and Baerbel Inhelder. *The Child's Conception of Space*. Norton, New York, 1967. (first published in French, 1948).

[Pierce and Kuipers 1991] David M. Pierce and Benjamin J. Kuipers. Learning hill-climbing functions as a strategy for generating behaviors in a mobile robot. In J.-A. Meyer and S. W. Wilson, editors, *From Animals to Animats: Proceedings of The First International Conference on Simulation of Adaptive Behavior (SAB-90)*, pages 327–336, Cambridge, MA, 1991. MIT Press/Bradford Books. (Also University of Texas at Austin, AI Laboratory TR AI91-137).

[Pierce 1991a] David Pierce. Learning a set of primitive actions with an uninterpreted sensorimotor apparatus. In *Proceedings of the Eighth International Workshop on Machine Learning*, pages 338–342. Morgan Kaufmann, 1991.

[Pierce 1991b] David M. Pierce. Learning turn and travel actions with an uninterpreted sensorimotor apparatus. In *Proceedings IEEE International Conference on Robotics and Automation*, pages 391–404, Hillsdale, NJ, April 1991. Lawrence Erlbaum Associates, Publishers.

[Pitt and Warmuth 1989] Leonard Pitt and Manfred K. Warmuth. The minimum consistent DFA problem cannot be approximated within any polynomial. In *Proceedings of the Twenty First Annual ACM Symposium on Theoretical Computing*, pages 421–432, 1989.

[Pomerleau *et al.* 1991] D.A. Pomerleau, J. Gowdy, and C.E. Thorpe. Combining artificial neural networks and symbolic processing for autonomous robot guidance. *Engineering Applications of Artificial Intelligence*, 4(4):279–285, 1991.

[Pomerleau 1989] D.A. Pomerleau. Alvinn: An autonomous land vehicle in a neural network. In D.S. Touretzky, editor, *Advances in Neural Information Processing Systems, 1*, San Mateo, CA, 1989. Morgan Kaufmann.

[Pomerleau 1990] D.A. Pomerleau. Neural network based autonomous navigation. In Charles Thorpe, editor, *Vision and Navigation: The CMU Navlab*. Kluwer Academic Publishers, Boston, MA, 1990.

[Quinlan 1983] J. Ross Quinlan. Learning efficient classification procedures and their application to chess end games. In Ryszard S. Michalski, Jaime G. Carbonell, and Tom M. Mitchell, editors, *Machine Learning, Volume 1*. Tioga, Palo Alto CA, 1983.

[Reid *et al.* 1981] L.D. Reid, E.N. Solowka, and A.M. Billing. A systematic study of driver steering behaviour. *Ergonomics*, 24:447–462, 1981.

[Rivest and Schapire 1987] Ronald L. Rivest and Robert E. Schapire. Diversity-based inference of finite automata. In *Proceedings of the Twenty Eighth Annual Symposium on Foundations of Computer Science*, pages 78–87, 1987.

[Rivest and Schapire 1989] Ronald L. Rivest and Robert E. Schapire. Inference of finite automata using homing sequences. In *Proceedings of the Twenty First Annual ACM Symposium on Theoretical Computing*, pages 411–420, 1989.

[Ross 1983] S. Ross. *Introduction to Stochastic Dynamic Programming*. Academic Press, New York, 1983.

[Rudrich 1985] Steven Rudrich. Inferring the structure of a Markov chain from its output. In *Proceedings of the Twenty Sixth Annual Symposium on Foundations of Computer Science*, pages 321–326, 1985.

[Rumelhart *et al.* 1986] D.E. Rumelhart, G.E. Hinton, and R.J. Williams. Learning internal representations by error propagation. In D.E. Rumelhart and J.L. McClelland, editors, *Parallel Distributed Processing: Explorations in the Microstructures of Cognition. Vol. 1: Foundations.* Bradford Books/MIT Press, Cambridge, MA, 1986.

[Sage and White 1977] A. P. Sage and C. C. White. *Optimum Systems Control.* Prentice Hall, 1977.

[Samuel 1959] A. L. Samuel. Some Studies in Machine Learning using the Game of Checkers. *IBM Journal on Research and Development*, 1959.

[Schapire 1991] Robert E. Schapire. The design and analysis of efficient learning algorithms. Technical Report TR-493, MIT Laboratory for Computer Science, 1991.

[Schmidhuber 1990] Jurgen Schmidhuber. Making the world differentiable: on using self-supervised fully recurrent neural networks for dynamic reinforcement learning and planning in non-stationary environments. Technical Report Report FKI-126-90 (revised), Technische Universitat Munchen, 1990.

[Schoppers 1987] Marcel J. Schoppers. Universal plans for reactive robots in unpredictable domains. In *Proceedings of IJCAI-87*, pages 1039–1046, 1987.

[Sedgewick 1983] Robert Sedgewick. *Algorithms.* Addison-Wesley, Reading MA, 1983.

[Segre 1988] Alberto Maria Segre. *Machine Learning of Assembly Plans.* Kluwer, Norwell MA, 1988.

[Shiller and Chen 1990] Zvi Shiller and J. C. Chen. Optimal motion planning of autonomous vehicles in three dimensional terrains. In *Proceedings IEEE International Conference on Robotics and Automation*, pages 198–203, 1990.

[Singh 1991] S. P. Singh. Transfer of learning across compositions of sequential tasks. In *Machine Learning: Proceedings of the Eighth International Workshop*. Morgan Kaufman, June 1991.

[Singh 1992] Satinder Singh. Transfer of learning by composing solutions of elemental sequential tasks. *Machine Learning*, 8:323–339, 1992.

[Smith and Cheeseman 1986] R. C. Smith and P. Cheeseman. On the representation and estimation of spatial uncertainty. *The International Journal of Robotics Research*, 5(4), 1986.

[Stansfield 1988] Sharon A. Stansfield. A robotic perceptual system utilizing passive vision and active touch. *International Journal of Robotics Research*, 7(6):138–161, December 1988.

[Sutton and Barto 1990] R. S. Sutton and A. G. Barto. Time-Derivative Models of Pavlovian Reinforcement. In M. Gabriel and J. Moore, editors, *Learning and Computational Neuroscience: Foundations of Adaptive Networks*. MIT Press, 1990.

[Sutton 1984] Richard S. Sutton. *Temporal Credit Assignment In Reinforcement Learning*. PhD thesis, University of Massachusetts at Amherst, 1984. (Also COINS Tech Report 84-02).

[Sutton 1988] R. S. Sutton. Learning to Predict by the Methods of Temporal Differences. *Machine Learning*, 3:9–44, 1988.

[Sutton 1990] R. S. Sutton. Integrated Architectures for Learning, Planning, and Reacting Based on Approximating Dynamic Programming. In *Proceedings of the 7th International Conference on Machine Learning*. Morgan Kaufman, June 1990.

[Sutton 1991] Richard S. Sutton. Planning by incremental dynamic programming. In *Proceedings of the Eighth International Workshop on Machine Learning*, pages 353–357. Morgan Kaufmann, 1991.

[Swarup 1991] Nitish Swarup. Q-maps: A neurological/visual approach to solving the structural credit assignment problem in reinforcement learning. Technical Report RC 17217, IBM Computer Science, 1991.

[Tan 1991] Ming Tan. *Cost Sensitive Robot Learning*. PhD thesis, Carnegie Mellon University, 1991.

[Tesauro 1991] G. J. Tesauro. Practical Issues in Temporal Difference Learning. RC 17223 (76307), IBM T. J. Watson Research Center, NY, 1991.

[Thrun 1992] Sebastian Thrun. Efficient exploration in reinforcement learning. Technical Report CMU-CS-92-102, School of Computer Science, Carnegie Mellon University, 1992.

[Turk et al. 1988] M. A. Turk, D. G. Morgenthaler, K. D. Gremban, and M. Marra. VITS – A vision system for autonomous land vehicle navigation. *IEEE Transactions on Pattern Analysis and Machine Intelligence*, 10, 1988.

[Tzeng 1992] Wen-Guey Tzeng. Learning probabilistic automata and Markov chains via queries. *Machine Learning*, 8:151–166, 1992.

[Valiant 1984] L. G. Valiant. A theory of the learnable. *Communications of the ACM*, 27:1134–1142, 1984.

[Vayda and Kak 1991] A. J. Vayda and A. C. Kak. A robot vision system for recognition of generic shaped objects. *CVGIP: Image Understanding*, 54(1):1–46, July 1991.

[Viola 1990] Paul A. Viola. Neurally inspired plasticity in oculomotor processes. In *IEEE Conference on Neural Information Processing Systems – Natural and Synthetic*, November 1990.

[Waibel et al. 1988] A. Waibel, T. Hanazawa, G. Hinton, K. Shikano, and K. Lang. Phoneme recognition: Neural networks vs. hidden Markov models. In *Proceedings of the International Conference on Acoustics, Speech and Signal Processing*, New York, NY, 1988.

[Wallace et al. 1985] R. Wallace, A. Stentz, C. Thorpe, H. Moravec, W. Whittaker, and T. Kanade. First results in robot road-following. In *Proceedings of International Joint Conference on Artificial Intelligence*, 1985.

[Walter 1987] S. Walter. The sonar ring: Obstacle detection for a mobile robot. In *Proceedings IEEE International Conference on Robotics and Automation*, pages 1574–1579, 1987.

[Watkins and Dayan 1992] Christopher J. C. H. Watkins and Peter Dayan. Technical note: Q-learning. *Machine Learning*, 8:279–292, May 1992. (special issue on reinforcement learning).

[Watkins 1989] C. J. C. H. Watkins. Learning from Delayed Rewards. PhD. Thesis, King's College, University of Cambridge, May 1989.

[Wehner 1987] Rüdiger Wehner. Matched filters – neural models of the external world. *Journal of Comparative Physiology*, 161:511–531, 1987.

[Wells III 1989] William M. Wells III. Visual estimation of 3-d line segments from motion – a mobile robot vision system. *IEEE Transactions on Robotics and Automation*, 5(6):820–825, 1989.

[Whitehead and Ballard 1989] Steven D. Whitehead and Dana H. Ballard. A role for anticipation in reactive systems that learn. In *Proceedings of the Sixth International Workshop on Machine Learning*, Ithaca, NY, 1989. Morgan Kaufmann.

[Whitehead and Ballard 1990] Steven Whitehead and Dana Ballard. Active perception and reinforcement learning. In *Proceedings of the Seventh International Conference on Machine Learning*, June 1990.

[Whitehead and Ballard 1991] Steven D. Whitehead and Dana H. Ballard. A study of cooperative mechanisms for faster reinforcement learning. TR 365, Computer Science Dept., University of Rochester, February 1991.

[Whitehead 1991a] Steven D. Whitehead. A complexity analysis of cooperative mechanisms in reinforcement learning. In *Proceedings of the Ninth National Conference on Artificial Intelligence (AAAI-91)*, pages 607–613, 1991. (A similar version also appears in the Proceedings of the Eighth International Workshop on Machine Learning, Evanston, IL, June 1991.).

[Whitehead 1991b] Steven D. Whitehead. *Reinforcement Learning for the Adaptive Control of Perception and Action*. PhD thesis, Department of Computer Science, University of Rochester, Rochester, NY, November 1991.

[Widrow 1962] Bernard Widrow. Generalization and information storage in networks of ADALINE neurons. In Marshall C. Yovits, George T. Jacobi, and Gordon D. Goldstein, editors, *Self-Organizing Systems*, pages 435–461. Spartan, Washington, 1962.

[Williams 1987] Ronald J. Williams. Reinforcement–learning connectionist systems. Technical Report NU-CCS-87-3, College of Computer Science, Northeastern University, Boston; MA, 1987.

[Yang 1992] Qiang Yang. A theory of conflict resolution in planning. *Artificial Intelligence*, 58:316–392, 1992.

[Zaharakis and Guez 1990] Steven C. Zaharakis and Allon Guez. Time optimal robot navigation via the slack set method. *IEEE Transactions on Systems, Man, and Cybernetics*, 20:1396–1407, December 1990.

INDEX